Lecture Notes in Artificial Intelligence 9577

Subseries of Lecture Notes in Computer Science

More information about this series at http://www.springer.com/series/1244

José F. Quesada · Francisco-Jesús Martín Mateos
Teresa Lopez-Soto (Eds.)

Future and Emergent Trends in Language Technology

First International Workshop, FETLT 2015
Seville, Spain, November 19–20, 2015
Revised Selected Papers

 Springer

Editors
José F. Quesada
University of Seville
Seville
Spain

Teresa Lopez-Soto
University of Seville
Seville
Spain

Francisco-Jesús Martín Mateos
University of Seville
Seville
Spain

ISSN 0302-9743   ISSN 1611-3349   (electronic)
Lecture Notes in Artificial Intelligence
ISBN 978-3-319-33499-8   ISBN 978-3-319-33500-1   (eBook)
DOI 10.1007/978-3-319-33500-1

Library of Congress Control Number: 2016936966

LNCS Sublibrary: SL7 – Artificial Intelligence

Printed on acid-free paper

This Springer imprint is published by Springer Nature
The registered company is Springer International Publishing AG Switzerland

Preface

This book constitutes the proceedings of the Workshop in Future and Emerging Trends in Language Technology 2105, held at the University of Seville, Spain, in November 2015.

The volume includes the abstract of the seven invited keynote speakers, three position papers, and the ten regular papers selected for presentation at the workshop after a peer-review process.

Language is one of the most distinctive human skills. Our ability to connect the mind and the world by means of language is simply amazing. Our competence to generate, process, and understand a theoretically infinite number of linguistic expressions involves many functional and cognitive capabilities. Information technologies have become so commonplace in our everyday lives, that it is almost impossible to think of any activity that does not require or involve the use of such technologies. And yet the use of information technologies to manipulate, understand, or generate different types of linguistic phenomena has been for decades, and continues to be, a crucial challenge.

Despite the aforementioned challenges and complexities around this field, we must highlight the outstanding body of knowledge, best practices, tools, and techniques described and delivered by researchers in language technology. Moreover, this field will have a crucial economic and social impact in the near future as some initiatives are showing. Among these initiatives, the multilingual single digital market is a relevant catalyst.

A persistent paradox appears on the application of language technologies to common and real application areas. As native speakers, we have a fluent and effortless high control of the language. By contrast, a machine is able to calculate, in less than a second, more mathematical operations that we (the speakers) would not be able to process even if we had an entire year to do so. Paradoxically, that same machine will have tremendous difficulties to manipulate the language used by a young 3-year-old girl. This creates great expectations about the capabilities of a machine to manipulate the language, while, at the same time, disappointment shows up when these expectations are not met adequately.

Definitely, language technology constitutes a mature research and development domain. However, many improvements must be addressed in order to obtain more robust and wider applications at an industrial and commercial level.

Taking into account this scenario, the FETLT 2015 workshop was planned, organized, and scheduled with the main goal of facilitating the interchange of ideas and networking of people around one key topic: "Future and Emerging Trends in Language Technology." This volume includes the material presented at the workshop, held during November 19–20, 2015.

The first part of this volume presents the abstracts of the seven invited presentations. During the opening session, Steve Young (Cambridge University) focused on one of the most prominent topics in the spoken dialogue systems domain: "Towards Open Domain Spoken Dialogue Systems," and Sebastian Möller (TU Berlin; Telekom Innovation Labs) concentrated on "Motivating New Interaction Experiences Involving Implicit Interaction, Body Sensors, Adaptive, and Persuasive Interface." The workshop was closed by Pierre-Paul Sondag (European Commission) talking about "Speech and Dialogue Technologies, Assets for the Multilingual Digital Single Market."

The workshop also scheduled a special session on research projects, with the participation of four invited speakers representing different EU-funded research projects: Asunción Gómez (Universidad Politécnica de Madrid) "Linguistic Linked Data: Paving the Way Towards Maximising (Re)Usability of Linguistic Resources"; Giuseppe Riccardi (University of Trento) "SENSEI: Making Sense of Human Conversations"; Núria Bel (Universitat Pompeu Fabra) "META-NET: Multilingual Europe Technology Alliance - Network of Excellence;" and Steve Renals (University of Edinburgh) "A Roadmap for Conversational Interaction Technologies."

Some of these keynote speakers contributed with position papers that are also included in this volume. Finally, these proceedings incorporate the ten regular papers accepted after the peer-review process carried out by the members of the Program Committee. Twenty proposals were received and acceptance was based solely on the evaluation of the referees and the achieved scores.

November 2015   Teresa Lopez-Soto
                                              Francisco-Jesús Martín-Mateos
                                                      José F. Quesada

Organization

FETLT 2015, the First International Conference on Future and Emerging Trends in Language Technology, was held at University of Seville, Seville, November 19–20, 2015.

Conference Direction

General chair

José F. Quesada   University of Seville, Spain

Program Committee and Advisory Group

Alex Acero	Apple, USA
Roberto Basili	University of Rome, Italy
Núria Bel	University Pompeu Fabra, Spain
Johan Bos	University of Groningen, The Netherlands
Nicoletta Calzolari	CNR-ILC, Italy
Khalid Choukri	ELDA, France
Walter Daelemans	University of Antwerp, Belgium
Thierry Declerck	DFKI, Germany
Marc Dymetman	Xerox Research Centre Europe, France
Antonio Ferrandez	University of Alicante, Spain
Ana García-Serrano	UNED, Spain
Jesús Giménez	Nuance Communications, USA
Xavier Gómez-Guinovart	University of Vigo, Spain
Gregory Grefenstette	Inria, France
Veronique Hoste	University of Ghent, Belgium
Eduard Hovy	Carnegie Mellon University, USA
Rebecca Jonson	Artificial Solutions, Sweden
Alon Lavie	Carnegie Mellon University, USA
Ramón López-Cózar	University of Granada, Spain
Teresa Lopez-Soto	University of Seville, Spain
Roberto Manione	AlliumTech, Italy
Daniel Marcu	University of Southern California, USA
Joseph Mariani	LIMSI-CNRS and IMMI, France
Patricio Martínez-Barco	University of Alicante, Spain
Ruslan Mitkov	University of Wolverhampton, UK
Antonio Moreno-Sandoval	Autonomous University of Madrid, Spain
Sergei Nirenburg	Rensselaer Poytechnic Institute, USA
José Manuel Pardo	Universidad Politécnica de Madrid, Spain
Mirko Plitt	Modula Language Automation, Switzerland

Massimo Poesio	University of Essex, UK; University of Trento, Italy
Andrei Popescu-Belis	Idiap Research Institute, Switzerland
José F. Quesada	University of Seville, Spain
Manny Rayner	University of Geneva, Switzerland
Steve Renals	University of Edinburgh, UK
Giuseppe Riccardi	University of Trento, Italy
Francisco J. Salguero	University of Seville, Spain
Kepa Sarasola	University of the Basque Country, Spain
Javier Sastre	Ateknea Solutions, Spain
Marc Steedman	University of Edinburgh, UK
David Suendermann-Oeft	Educational Testing Service, USA
Khiet Truong	University of Twente, The Netherlands
Alfonso Ureña	University of Jaen, Spain
Jason D. Williams	Microsoft Research, USA

Organizing Committee

Joaquín Borrego-Díaz	University of Seville, Spain
Juan Galán-Páez	University of Seville, Spain
Diego Jiménez	University of Seville, Spain
Teresa Lopez-Soto	University of Seville, Spain
Francisco-Jesús Martín-Mateos	University of Seville, Spain
Ángel Nepomuceno	University of Seville, Spain
José F. Quesada	University of Seville, Spain
Francisco J. Salguero	University of Seville, Spain

Invited Speakers

Towards Open Domain Spoken Dialogue Systems

Steve Young

Cambridge University

The number of networked IT related devices is increasing rapidly and each has its own interface, increasingly constrained by the physical dimensions of the device. Speech has the potential to provide a single uniform interface to all our IT services and devices but to be effective, speech interfaces must provide full dialogue capability not just simple spoken commands. This talk will discuss the current state of the art in spoken dialogue systems and discuss the issues involved in scaling from today's limited domain systems to fully open domains.

Motivating New Interaction Experiences Involving Implicit Interaction, Body Sensors, Adaptive, and Persuasive Interfaces

Sebastian Möller

TU Berlin; Telekom Innovation Labs

In most cases, speech and language technology is used for explicit interaction with computers. However, in a connected world the technology can also to be used for other purposes. It is the aim of this position statement to open research questions which point at the potential of speech and language technology for enabling new interaction experiences such as implicit interactions, body sensors, adaptive and persuasive interfaces. The research questions are grouped into three domains: user, system and context. In the user domain, our focus will be on information that can be extracted from the user. In the system domain, our focus in on crowdsourcing and privacy concerns. The focus point in the context domain is on implicit interactions and intelligent interactions.

Speech and Dialogue Technologies, Assets for the Multilingual Digital Single Market

Pierre-Paul Sondag

European Commission

In the early days machine translation and speech recognition or generation, were developed as two separate strands of technologies, keeping machine translation, dialogue handling and speech processing isolated as autonomous systems dedicated for one single task. Over time, these different technologies converged as they became all data driven, while in the same, speech and language technologies had to be integrated into complex systems in order to overcome the challenges of interaction between humans and machines. Today human-machine or computer mediated human to human dialogues systems combine language technologies, speech processing and advanced semantics to allow more natural and more spontaneous ways of dialogues. The dialogue module became the glue that brought together and intertwined these technologies, increasing their performance and usability up to a level appropriate for the needs of real world applications. Speech, particularly when enhanced with other modalities, remains the most common and natural way to interact. Applied together with localisation and machine translation, speech will provide access for all people, including the less computer literate, to digital services, and hence ease the advent of the multilingual Digital Single Market.

Linguistic Linked Data: Paving the Way Towards Maximising (Re)Usability of Linguistic Resources

Asunción Gómez

Universidad Politécnica de Madrid

Language is one of the most important cultural assets of mankind. Accordingly, the study of language has occupied and is still occupying countless researchers worldwide who study phenomena related to phonological, morphological, syntactic, semantic and pragmatic aspects of languages. Most linguistic research is conducted by empirically observing the competence and behaviour of language users and thus requires access to actual linguistic data. A linguistic data ecosystem is therefore needed by many stakeholders, including: (i) linguists, (ii) translators, (iii) terminologists, (iv) Natural Language Processing (NLP) software developers that need to train their system on linguistic data sets. Such an ecosystem would thus have a tremendous potential impact. As of today, however, there is no mature holistic ecosystem that would foster the systematic discovery, exploration, exploitation, extension, curation and quality control of linguistic data, due to heterogeneous formats and annotation standards.

The Linked Data (LD) paradigm that has recently emerged provides a suitable approach for the development of such an ecosystem, supporting: data set discovery with relevance for a particular task, interoperability across heterogeneous annotation formats, composition and integration of datasets to make analyses and statistics more robust, compliance to existing policies and terms and conditions, Web accessibility and standardization, Semantic Web based querying and data management, quality assurance and benchmarking as well as diagnosis and repair procedures, collaborative and distributed data set evolution and curation, among others.

LD refers to the recommended best practices for exposing, sharing, and connecting data on the Web. LD is multilingual. LD builds in particular on RDF as a data model for representing structured content. RDF is a directed graph model where resources, identified by URIs, are given attributes and connected to other resources by means of properties, also identified by URIs, thus forming a structured graph that can be queried via the SPARQL query language. In LD, properties (links) can be navigated in a manner similar to the way we navigate HTML hyperlinks using a web browser. Clients can navigate from one RDF resource to another by dereferencing the URIs of related (linked) resources. It is important to emphasize that there is a clear distinction between URIs as unique identifiers of entities, concepts etc. – referring to language-independent entities existing in the real world – and the language symbols (i.e., labels) by which these entities are referred to. LD is implemented using standard Web protocols and dereferencing mechanisms, in particular content negotiation to request RDF content. Due to this fact, the LD cloud is inherently associated with human-readable descriptions (mainly in HTML) of the resources described in RDF. This implies that RDF and textual (HTML) content do not just live next to each other on the Web of Data, but are also indirectly connected to each other.

The key benefits of applying LD principles to linguistic data are (i) better modelling of datasets as directed labelled graphs, (ii) structural interoperability of heterogeneous resources, (iii) federation of resources from different sources and at different layers of linguistic annotation, (iv) a strong ecosystem of tools based on RDF and SPARQL, (v) improved conceptual interoperability due to strong semantic models such as OWL and shared semantics due to linking and (vi) dynamic evolution of resources on the web. Linguistic Linked Data (LLD) aims at applying LD principles to the publication, curation, exploration and use of Linguistic Data. In this talk I will explore the concepts of Linked Data, multilingual Linked Data and Linguistic Linked DAta.

SENSEI: Making Sense of Human Conversations

Giuseppe Riccardi

University of Trento

Conversational interaction is the most natural and persistent paradigm for business relations with customers. In contact centres millions of calls are handled daily. On social media platforms millions of blog posts are exchanged amongst users.

Can we make sense of such conversations and help create assets and value for private and public organizations' decision makers? And indeed for anyone interested in conversational content?

The overall goals of the SENSEI project are twofold. First, SENSEI will develop summarization/analytics technology to help users make sense of human conversation streams from diverse media channels. Second, SENSEI will design and evaluate its summarization technology in real-world environments, aiming to improve task performance and productivity of end-users.

META-NET: Multilingual Europe Technology Alliance - Network of Excellence

Núria Bel

Universitat Pompeu Fabra

META-NET is a Network of Excellence dedicated to fostering the technological foundations of a multilingual European information society. Language Technologies will:

- Enable communication and cooperation across languages.
- Secure users of any language equal access to information and knowledge.
- Build upon and advance functionalities of networked information technology.

A concerted, substantial, continent-wide effort in language technology research and engineering is needed for realising applications that enable automatic translation, multilingual information and knowledge management and content production across all European languages. This effort will also enhance the development of intuitive language-based interfaces to technology ranging from household electronics, machinery and vehicles to computers and robots.

To this end META-NET is building the Multilingual Europe Technology Alliance (META). Bringing together researchers, commercial technology providers, private and corporate language technology users, language professionals and other information society stakeholders. META will prepare the necessary ambitious joint effort towards furthering language technologies as a means towards realising the vision of a Europe united as one single digital market and information space.

A Roadmap for Conversational Interaction Technologies

Steve Renals

University of Edinburg

ROCKIT is a European project to construct a technology roadmap for Conversational Interaction Technologies, which is concerned with technologies for multilingual Human-Human, Human-Machine, and Human-Environment interactions. The underlying vision is for technologies which support natural communication and are multimodal, multidevice, and transferable across domains. The roadmap conveys the relationships among societal drivers of change, products and services, use cases for them, and research results, and aims to bridge research and innovation. In this talk I'll discuss the ROCKIT roadmapping process, which has involved a wide range of stakeholders, and its outputs including a set of target research and innovation scenarios.

Contents

Position Papers

Motivating New Interaction Experiences Involving Implicit Interaction, Body Sensors, Adaptive, and Persuasive Interfaces

Sebastian Möller[✉] and Jan-Niklas Antons

Quality and Usability Lab, Telekom Innovation Laboratories,
TU Berlin, Berlin, Germany
{sebastian.moeller,jan-niklas.antons}@telekom.de

Abstract. In most cases, speech and language technology is used for explicit interaction with computers. However, in a connected world the technology can also to be used for other purposes. It is the aim of this position statement to open research questions which point at the potential of speech and language technology for enabling new interaction experiences such as implicit interactions, body sensors, adaptive and persuasive interfaces. The research questions are grouped into three domains: user, system and context. In the user domain, our focus will be on information that can be extracted from the user. In the system domain, our focus in on crowdsourcing and privacy concerns. The focus point in the context domain is on implicit interactions and intelligent interactions.

Keywords: Interaction experiences · Implicit interaction · Body sensors · Adaptive interfaces · Persuasive interfaces

1 Introduction

Speech and language technology has been designed in the past as an explicit interaction technique, mainly for replacing windows, icons, menus and pointing device (WIMP) interfaces in some applications. The underlying assumption is that humans use computers in a dedicated way to perform a pre-defined task. Thus, the interaction requires fully-capable humans who have no other activities than interacting with a concrete interface to do something they are fully aware of. While there are exceptions to this rule, most of today's speech-based applications (dictation systems, telephone-based spoken dialogue systems, etc.) follow this paradigm. Exceptions to this rule are e.g. speech-based interaction with a navigation system in a car, speech interfaces for the blind, or adaptive systems which integrate the user state in the adaptation strategy and can therefore compensate high workload in the main task.

However, in a connected world, ubiquitous computing capabilities and mobile sensing devices allow speech and language technologies to be used for other

© Springer International Publishing Switzerland 2016
J.F. Quesada et al. (Eds.): FETLT 2015, LNAI 9577, pp. 3–9, 2016.
DOI: 10.1007/978-3-319-33500-1_1

purposes as well: Speech as the most important biosignal permits to continuously sense information about the user; spoken language interfaces integrated into smart environments or accessed through mobile devices allow speech interaction in nearly all everyday situations; low-cost and widespread availability of the respective technology allows it to be used by under-represented groups of people, and with special needs; and crowdsourcing platforms allow data to be handled with human intelligence to a degree which was rarely possibly in the past, building the basis for new robust applications. These changes open speech and language technologies for applications which are hardly conceived and explored today.

It is the aim of this position statement to open research questions which point at the potential of speech and language technology for enabling new interaction experiences. The questions are not meant to be complete or conclusive, and no answers will be given to them in this paper. Rather, we will try to illustrate their relevance through exemplary applications which would take profit of such technologies. By this, we hope to motivate research into selected areas which we consider substantial for progress towards such applications. This research will not necessarily address technological challenges (which are undoubtly present), but rather the user's perspective on the offered technologies. Such a perspective will prioritize research from a user-relevance consideration.

In the following sections, we will try to group our research questions into three domains:

– User domain
– System domain
– Context domain.

For each domain, we will explain why we consider the questions to be relevant, and what type of research is necessary to explore or answer these questions. We will also provide exemplary applications as illustrations of our points.

2 User Domain

We consider the user as a primary source of information for each speech- or language-based interaction. Thus, identifying user characteristics (including general user characteristics such as their cognitive state), understanding user motivations and needs and tracking their experience are a target of paramount importance. The following research questions would need to be answered:

– *Which information can be extracted from the user?*
  In the past, we have seen that the user's age, gender, affective state, fatigue, level of intoxication, and other paralinguistic characteristics can be extracted from speech. The series of challenges organized at INTERSPEECH conferences offers a good illustration of the current state-of-the-art in this respect. Consequently, it could be shown that these user factors cannot only be extracted from speech, but also have a significantly influence on the user's

experience [1]. However, speech is only one of several biosignals which modern sensors built in e.g. mobile devices are able to capture. Fusing information from different sensors, and aggregating them into meaningful states, would offer potential of adapting new services to the current user state.

An example is personality recognition from speech, which has proven to be feasible in several circumstances [7]. When being able to recognize a user's personality (e.g. in terms of the "big five" traits), and combining this information with predictions of the user's affective state (extracted e.g. from EEG sensors fixed to a headset), this would allow to proactively adapt interactive services according to the expected relevance for the user.

- *What motivates users for specific interactions?*

Understanding the user's needs, motivations, and states will help to design interactions which are relevant, and thus better fulfill their requirements. Unfortunately, little is known about why users perform specific interactions and omit others. Motivational theories are mostly adapted towards professional work motivation, and less so to on-the-side activities.

We expect that a thorough understanding of a specific user's motivations will help to design better interaction techniques. A prerequisite to this are empirical studies on motivation, and psychometric tools for analyzing the major motivational components. Then, strategies for need fulfillment have to be developed which should take into account environmental and other contextual factors (see below). The success of these strategies should be carefully tracked, in the best case by optimally adapting interfaces.

An example would be a speech-based personal assistant. This assistant would not simply answer a user's questions, but try to find the underlying need of a request. It would then develop strategies for fulfilling this need in the best possible way, using speech or other interaction modalities, whatever best fulfills the purpose. The success of these strategies would be continuously monitored as to whether they lead to positive experiences in the user.

A second example would be adaptive software for cognitive training, such as the one developed in the PflegeTab project at TU Berlin. Although most software is mainly focused on graphical user interfaces, many users with special needs will benefit by focusing on speech as an interaction modality. For example, users with dementia often have sensory impairments in the visual domain. An adaptation to the user's characteristics can not only provide instruction on what a given task is, but also on how to solve a given task. As the matching between the capabilities of the user and the demand of the task is met, a positive influence on the motivation and satisfaction is assumed.

- *How can body-related interfaces be used for new interaction modalities?*

As indicated above, speech- or natural-language-based interaction might not be the optimum one in all situations. We could imagine silent speech interfaces for situations where a user would like to confidentially communicate with one of several present interlocutors, or where parents might want to communicate confidentially without awareness of their children present on the spot. Sensor and actuator techniques for this purpose could be integrated into wearables to allow for socially acceptable interactions in the presence of others.

Examples for non-speech based silent interaction technique are MagiThings and NeuroPad. MagiThings enables gestural interactions with mobile devices based on using embedded compass (magnetic field) sensors [4]. Whereas NeuroPad is an iPad-application that connects a commercially available low-cost neuro headset with an iPad [5]. The extracted physiological signals are used for controlling different functionalities in a touchless manner.

– *How can user experience be tracked?*
A consequent user-driven approach to interaction design requires that the effect of the interaction on the user is known and can be measured. Unfortunately, present techniques for measuring user experience are mostly questionnaire-based tools, which are difficult to apply in an interactive situation, mostly refer to a retrospective experience, and partially destroy the experience they are intended to measure. As an alternative, physiological measures such as electroencephalogram (EEG), heart rate variability, skin conductance, or alike have shown to indicate quality of experience and user state, see [2,3]. These measures should be explored further for interactive situations, as they could fit into an experience feedback loop for adaptive interactive systems.

3   System Domain

There are a multitude of system components which would need further research and development in order to fit into the paradigms illustrated above. As we cannot cover all of these components and technologies, we prefer to just raise two fundamental questions related to data-driven speech and language technology:

– *How can speech and multimodal user data be collected through crowdsourcing platforms?*
Crowdsourcing has proven to be an important data collection technique for many applications, and several companies build their business model on such techniques. In speech and language technology research, however, this has so far seen less attraction. This is particularly astonishing, since the availability of workers collecting data in different locations, language areas, environmental conditions, etc. would help to make a leap towards real-life and diverse databases which could significantly improve system robustness. A frequently-heard reason for the non-interest is the assumedly low data quality. Still, research on the quality of crowdsourcing, and especially on factors impacting it, is sparse [6]. It is mostly unknown why workers get attached to a platform and accept certain jobs, and what is on the source of their motivation.

Thus, we expect research to be necessary analyzing speech and language data collection in crowdsourcing environments. We foresee that mobile platforms, which would be able to collect data under diverse environmental conditions, are especially interesting. It will be necessary to investigate mechanisms of quality control for data which has no apparent ground truth. These techniques need to take user motivational factors and needs into account, see the discussion above.

– *How can speech and other biometric user data be collected and processed responsibly?*
As speech is a biosignal, it always leaves traces to its producer. With the growing proportion of data-driven approaches, and the growing number of databases addressing different types of information from the same (or overlapping) users, the question arises as to how data needs to be treated in order to respect the user's need for privacy. In principle, it would be possible to combine large speech corpora with other corpora of behavioral data (such as movement patterns, preferences, etc.), and in this way construct a detailed picture of the situation of a specific speaker. This would raise a fundamental concern for any individual who insists on self-determined use of their data.
Research on the use of information which can be created by combining different data sources should go hand-in-hand with user requirement analyses on security and privacy needs. This should build the basis for system architectures which are able to treat the different sources of data responsibly, and thus help to fulfill the user's needs. A careful consideration of data security and privacy should also keep the user in the loop, as their behavior will be decisive for the usefulness of the developed strategies.

4 Context Domain

As mentioned in the introduction, we expect that interactions with speech and language technologies will more and more become "implicit", being nested in or combined with other activities, without the situation of being confronted to a concrete interface. The following research questions might arise from these situations:

– *How can interactions take place without interfaces?*
In most speech-based applications, an interaction (or communication) takes place between two or more agents. These agents are concretized by their voice, their location, their apparent personality, etc. This paradigm might get blurred when speech is used as an interaction technique in so-called "intelligent environments". Such environments usually possess a multitude of different sensors and actuators, which follow different interaction and behavioral logics.
The question is how such interactions would take place, and following which (conversational) rules. Answering this question first requires a formalization and empirical analysis of human multimodal interaction behavior in such environments. Then, interaction patterns can be inferred and used as a basis for dialogue management. It is expected that such interactions differ from standard speech-based human-computer interactions which can be observed e.g. in telephone-based spoken dialogue services. The reasons underlying such differences might uncover new interaction paradigms, which are more efficient and perhaps also more fun-to-use and more motivating for users.
– *How do non-task-directed interactions take place?*
A particularly interesting class of interactions addresses situations which lack

a concrete, narrow task goal. Instead, the interaction is carried out in order to fulfill a user's need, and/or to reach a high-level goal. Such interactions could be part of persuasive interfaces, inciting the user to specific intended behavior (such as behaving energy-efficient or healthy), or part of an entertaining game. Designing interactions for such applications is particularly challenging, as they need to ground on the user's needs and motivational forces in order to reach their aim. As speech can be a particularly personalized and persuasive interaction technique, it might be particularly suited for this purpose. Strategies of gamification and adaptation could help to reach the target, although the principles underlying such strategies are not yet fully understood.

– *What makes an interaction intelligent?*
Interactions in networked environments, as well as interactions with persuasive interfaces, are frequently considered as being more "intelligent", without specifying what is actually meant by this term. In fact, knowledge is missing about what makes an interaction – and the responsible interaction partner – seem "intelligent" to the human interaction partner. Answering this question first requires a common understanding of, and measurement tools for, perceived intelligence, including negative counterparts such as reactance, which might lead to disappointed users. Intelligence should be distinguished from adaptivity; however, it is also of major importance to find out about the perceived intelligence of different adaptation strategies. One interesting question which arises in this context is if adaptive system which would probably objectively considered as more intelligent are perceived as such by the user. If the interventions by an adaptive system are not perceived by the user at all, the interaction flow would stay intact. The knowledge on intelligent interactions would help to build interactive systems which can be suited to the particular capabilities and needs of different users.

5   Conclusions

The list of questions raised in this position statement may appear vast, and there will be for sure progress in speech and language technology without answering any of these questions. However, we think that a user-driven perspective on speech and language technologies will be helpful in two ways. First, it will help to develop more relevant applications, which will have a higher impact on the market, and subsequently motivate further interest (and investment) into this field. Second, and perhaps more importantly, it will also improve our understanding on humans interacting with machines. Such an improved understanding will be helpful in many different ways, for many other technologies and applications alike.

References

1. Antons, J.N., Arndt, S., De Moor, K., Zander, S.: Impact of perceived quality and other influencing factors on emotional video experience. In: 2015 Seventh International Workshop on Quality of Multimedia Experience (QoMEX), pp. 1–6. IEEE (2015)

2. Antons, J.-N.: Neural Correlates of Quality Perception for Complex Speech Signals. Springer, Switzerland (2015)
3. Arndt, S.: Neural Correlates of Quality During Perception of Audiovisual Stimuli. Doctoral dissertation, TU Berlin (2015)
4. Ketabdar, H., Haji-Abolhassani, A., Roshandel, M.: MagiThings: gestural interaction with mobile devices based on using embedded compass (magnetic field) sensor. Int. J. Mob. Hum. Comput. Interact. (IJMHCI) 5(3), 23–41 (2013)
5. Matthies, D.J., Antons, J.-N., Heidmann, F., Wettach, R., Schleicher, R.: NeuroPad: use cases for a mobile physiological interface. In: Proceedings of the 7th Nordic Conference on Human-Computer Interaction: Making Sense Through Design, pp. 795–796. ACM (2012)
6. Naderi, B., Wechsung, I., Möller, S.: Effect of being observed on the reliability of responses in crowdsourcing micro-task platforms. In: 2015 Seventh International Workshop on Quality of Multimedia Experience (QoMEX), pp. 1–2. IEEE (2015)
7. Polzehl, T.: Personality in Speech. Springer, Switzerland (2015)

The SENSEI Project: Making Sense
of Human Conversations

Giuseppe Riccardi[1]($\boxtimes$), Frederic Bechet[2], Morena Danieli[1], Benoit Favre[2],
Robert Gaizauskas[3], Udo Kruschwitz[4], and Massimo Poesio[4]

[1] University of Trento, Trento, Italy
giuseppe.riccardi@unitn.it
[2] Aix-Marseille University, Marseille, France
[3] University of Sheffield, Sheffield, UK
[4] University of Essex, Colchester, UK
http://www.sensei-conversation.eu

Abstract. Conversational interaction is the most natural and persistent paradigm for personal and business relations. In contact centres customer spoken conversations are handled daily. On social media platforms conversations are delivered in different forms, lengths and for different purposes. In both cases, conversations have little impact on the intended target listeners, due to the volume, velocity and diversity (media, style, social context) of the document streams (spoken conversations and blog posts). Most language analytics technology is limited in that it performs keyword search, which does not provide automatic descriptions of what happened, who said what, which opinions are held on what subject, in a coherent, readable and executable form. In the SENSEI project we plan to go beyond keyword search and sentence-based analysis of conversations. We adapt lightweight and large coverage linguistic models of semantic and discourse resources to learn a layered model of conversations. SENSEI addresses the issue of multidimensional textual, spoken and metadata descriptors in terms of semantic, para-semantic and discourse structures. Automated generation of readable analytics documents (summaries) will support end-users in the context of large data analysis tasks. Summarization technology developed in SENSEI has been evaluated with respect to users' task requirements and performances in the context of contact centre and social media conversations.

Keywords: Summarization · Spoken dialogue · Social media · Language analytics

1 Introduction

Conversational interaction is the most natural and persistent paradigm for personal and business relations. Vast amounts of data of this type are already available to business, yet current language analytics technology only offers limited support. Data analysts facing such a data deluge, need to be able to extract and

© Springer International Publishing Switzerland 2016
J.F. Quesada et al. (Eds.): FETLT 2015, LNAI 9577, pp. 10–33, 2016.
DOI: 10.1007/978-3-319-33500-1_2

summarize relevant information from large quantities of this most fundamental form of human linguistic behaviour. For example, in contact centres millions of spoken conversations are handled daily to provide vital support to business units and their customers. However, a call centre analyst aiming to identify areas for improvement by examining the data collected by her/his company will only be able to study a tiny fraction of such data due to the limitations of speech analytics technology. Similar problems limit the analysis of comment threads on social media platforms, a new type of multiparty conversation in which hundreds of millions of blog posts and related comments are generated both in generalist (e.g. Twitter) or proprietary platforms (e.g. news websites). A journalist wanting to engage with his/her readers by following such threads will be quickly overwhelmed by the amount of data produced. Both types of conversations have limited impact on the intended target listeners due to the volume, velocity and diversity (media, style, social context) of the document streams (spoken conversations and blog posts). The SENSEI vision is to drive forward conversation analytics technology by addressing the state-of-the-art limitations, i.e. to develop analytics technologies that (1) understand conversations at a much deeper level, in particular taking account of para-semantic aspects of conversation (2) automatically generate a range of summary outputs to suit the range of end-users with a stake (e.g. conversation analysts) in making sense of large volumes of conversational data (3) are adaptable to different conversational channels and different user tasks.

This is a project review paper and we are going to refer to available studies and results we have achieved at this time and point to the companion website, [36], where the resources, including data, papers, use case design and reports are made available as they are published.

In the following section we will present the SENSEI vision regarding the modeling of summaries in two use case scenarios: (a) contact centre spoken conversations and (b) social media conversations. In Sect. 3 we review the parsing challenges, objectives and recent novel research work and experiments. In Sect. 4 we propose and motivate the conversation summary types in the context of dyadic spoken conversations and multi-party conversations generated on on-line social media platforms. In Sect. 5 we discuss summary evaluation scenarios for the two use cases.

2 Human Conversations

SENSEI's scientific and technology vision is motivated by both an ecological evaluation and the end-user task requirements. Ecological approaches to system evaluation include both observation of data generated by real industry processes as well as real end-user engagement. This is in contrast to largely unsuccessful top-down approaches that push niche and/or early-development technology into the development pipeline. To this end SENSEI has identified two use cases that are prime exemplars of the diverse space of applications for conversation analytics in the consolidated telephony and social media platforms. The two use cases pose similar technological challenges in terms of language understanding technology in real-world contexts. However such conversations occur over significantly different media (speech vs text) and social context (dyadic real-time conversations vs n-adic non-real time conversations). For each use case we have defined summary categories that we will propose and discuss in Sect. 4. Such summary categories may cover existing document types as well as new types that will address limitations of current analytics technology. Last but not least, the two use cases will allow us to instantiate both multimedia and cross-media investigation and technology development.

Call Centre Use Case. In outsourced call centres, large corporations outsource their customer touch-point to a hosting call centre. The in-coming and outgoing calls may be monitored in real time, or recorded for later review. The monitoring is done by human evaluators for small random call samples (much less than 1 %). Their job is to track indicators of call quality and agent efficiency. The call centre's corporate client may require reporting in different aggregated forms according to, e.g., the topic of the calls or, in other words, what their customers are asking about, or the emotional content of the call, e.g. concerned or frustrated user. The services provided by the human analysts and evaluators are very expensive in some cases or not feasible in others because of the data deluge or task complexity. The end-users of SENSEI analytics results are professional analysts working in call centres. Depending on the target of their evaluation (e.g. monitoring of agent efficiency, control of call quality, identification of call topic, evaluation of user satisfaction, evaluation of agent training needs), they will be able to profit from the different categories of summaries and reports generated by SENSEI systems.

Social Media Use Case. In a news publisher website such as *The Guardian* or *Le Monde*, journalists publish articles on different topics from politics and civil rights to health, sports and celebrity news. The website design supports the publication and consumption of original news articles and at the same time facilitates user-involvement via reader comments. Increasingly, in a period of disruptive change for the traditional media, newspapers see their future as lying in such conversations with and between readers, and new technologies to support these conversations will become essential. In this scenario there are a number of potential users:

- news readers and the originating journalist want to gain a structured overview of the mass of comments, both in terms of the sub-topics they address and their connection with the original article and in terms of the opinions (polarity and strength) the commenters hold about these topics;
- news readers who join a forum discussion need to be empowered so that they can respond to the originating article and/or to a sub-set of earlier comments that may be relevant to their own personal view on the matter;
- editors or media analysts may need a more widely scoping analysis.

At present none of these users can effectively exploit the mass of comment data – frequently hundreds of comments per article – as there are no tools to support them in doing so. What they need is new tools to help them make sense of this data deluge. In this scenario, therefore, SENSEI end-users will be news comment readers, news comment authors, journalists and editors/media analysts. Users in these categories will benefit from the various types of summaries and reports generated by SENSEI systems.

Figure 1 shows the overall architecture of SENSEI workflow in the context of the two use case scenarios. SENSEI conversational data are taken from call centres and social media platforms. They are parsed and annotated with semantic, para-semantic and discourse level descriptors and aggregated to yield summaries for end-users in the form of conversational-oriented summaries (e.g. topics T_i categorized using domain ontologies or multimedia extractive summaries), blogger-oriented summaries describing groups (e.g. group, G_i, orientations towards topic T_i), user-defined ad hoc reports (e.g. composition of semantic and para-semantic aspects) and rated questionnaires (e.g. call quality monitoring forms).

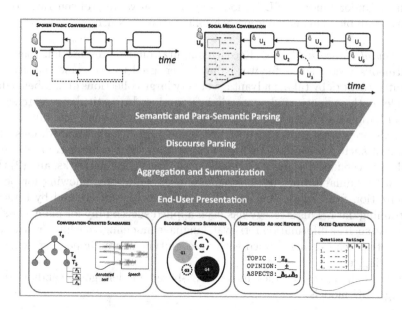

Fig. 1. SENSEI conversational analysis, parsing and summarization work-flow. Conversations are automatically annotated with semantic, para-semantic, discourse level descriptors and aggregated to yield summaries for end-users. The summaries are in the form of conversational-oriented summaries (e.g. topics T_i categorized using domain ontologies or multimedia extractive summaries), blogger-oriented summaries describing groups (e.g. group (G_i) orientations towards topic T_i), user-defined ad hoc reports (e.g. composition of semantic and para-semantic aspects) and rated questionnaires (e.g. call quality monitoring forms)

3 Parsing Human Conversations

3.1 Semantic Parsing

Semantic parsing is the process of producing semantic interpretations from words and other linguistic events that are automatically detected in a text conversation or a speech signal. Many semantic models have been proposed, ranging from formal models encoding deep semantic structures to shallow ones considering only the main topic of a document and the concepts or entities occurring in it. For Open Domain Semantic Parsing, generic purpose semantic models can be used, such as FrameNet or Abstract Meaning Representation (AMR). Once this generic meaning representation is obtained, a translation process trained on a small annotated corpus can be applied for projecting generic predicates and concepts to application specific ones. This kind of approach can help to reduce the need for large application-specific annotated corpora for training Natural Language Understanding (NLU) models by taking advantage of generic resources already available. This is the approach followed in SENSEI.

Deep Neural Network Models. Recent computational representations based on a continuous vector space for words have been used to overcome the need for annotated corpora by taking advantage of very large collections of unlabeled data to model both semantic and syntactic information. In particular researchers in Natural Language Processing have focused on learning a dense low dimensional (hundreds) representation space of words [38,47,53], called embeddings. The benefits of such representations are (1) that they offer a lower computational complexity when used as input of classifiers such as neural networks, and (2) that words with similar properties have similar representations, allowing for better generalization from subsequent models, e.g. for words not covered by targeted task training data. This strategy has been applied successfully for many classical NLP tasks such as information retrieval, language modeling, machine translation, as part-of-speech tagging, named entity recognition, syntactic parsing, semantic role labeling, etc.

Three main characteristics make DNN-based models good candidates for building NLU models:

- the use of a large amount of unlabeled data for learning word representations when dealing with a limited amount of in-domain data [58];
- the joint optimization of DNN over several NLP tasks;
- the ability of Recurrent Neural Networks (RNN) to maintain contextual information through sequence decoding with a memory model such as the Long Short Term Memory model [55].

This last characteristic is particularly relevant to SENSEI as one of its main foci is on the representation of conversational context in semantic parsing models.

One drawback of embeddings and DNN for semantic parsing on conversational data, as noted by [37], is the fact that they are usually obtained on very large written text corpora covering generic domains, such as news articles or

Wikipedia pages, although semantic parsing systems are dealing with spontaneous speech and non-canonical text on specific domains. To overcome this limitation in SENSEI we have proposed several adaptation methods along three dimensions: cross-media, cross-domain, cross-language.

These adaptation methods in SENSEI follow a common strategy:

- Open domain Semantic Parsing with generic semantic models such as Frame-Net.
- Joint use of a large amount of unlabeled data as well as rich linguistic resources in word embedding representation approaches when dealing with no or little in-domain annotated data.
- Adaptation to a new media/domain/language in the embedding space thanks to little adaptation data.

For example in [56] we address the cross-media/cross-domain issues by both adapting an embedding space trained on Wikipedia thanks to a small adaptation corpus containing spoken transcriptions corresponding to the call-centre we were dealing with; then by generalizing this adaptation to all words of the original embedding space, in particular to those not occurring in the adaptation corpus. A comparison of CRF and Neural Network methods is given in Fig. 2 for the semantic frame tagging task on the SENSEI call-centre corpus. This figure presents the results obtained by increasing the amount of adaptation data. CRF and NN only use word features. CRF++ uses as well Part-Of-Speech features; NN+ correspond to the adaptation process proposed in SENSEI.

3.2   Para-Semantic Parsing

Para-semantic parsing aims at analyzing paralinguistic features of human conversations and complements the semantic analysis of a conversation. Such features include turn-taking descriptors (e.g. speech overlap), speech rate, speech quality and pitch segmental statistics for spoken data and non-verbal cues such as text format features and emoticons for social media data. In SENSEI our goal is to investigate the relation with semantic features and aim at a joint or composite model. In social media analysis, most of the previous work on para-semantic traits has been done in the framework of Sentiment Classification, further divided into Opinion Detection and Sentiment Polarity Classification. An opinion can be defined as a quadruple: author (opinion holder), target audience, an object of the opinion, and semantic orientation (polarity) of the opinion (optionally also intensity of sentiment). The main focus of sentiment analysis research has been user reviews, to a much lesser degree blogs and forums, and significantly less dyadic or multiparty conversations. Thus, the analysis is generally limited to identification of semantic orientation, where supervised machine learning with bag-of-words models yields satisfactory performance. In the analysis of conversations, opinion holders, target audience and objects of opinion play a crucial role. Notable exceptions in the field are works that do stance classification in online debates or dialogues [Somasundaran and Wiebe, 2009, 2010]; they show

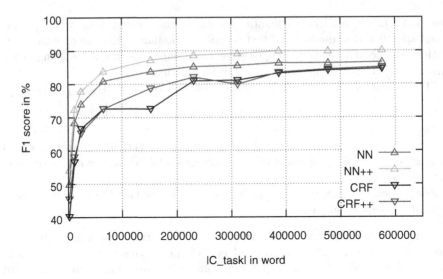

Fig. 2. Semantic Frame tagging performance (F-score) as function of the increasing amount of adaptation data. Comparison of CRF and baseline Neural Network approaches are shown as well the adaptation process proposed in SENSEI, denoted NN++.

that sentiment analysis of conversations requires a richer set of features, such as dialogue acts and discourse-based features. However, even in these works the full potential of discourse analysis is not explored, e.g. only discourse connectives are considered.

In spoken conversations in the last twenty years there has been a growing interest in and research work on *affective computing*, a comprehensive term including research on computational models of emotions, affect, personality and attitudes. However the analysis of emotions and computational models of them has been done in isolation from the semantic or discourse descriptions of human conversation. Last but not least, the emotion space (e.g. Ekman categories) is limiting for the richness and diversity of human conversations observed in-the-wild, such as public forums, business and personal communications.

Affective Scenes. In SENSEI, we have introduced the concept of affective scenes [42]. An affective scene is an emotional episode where one individual is affected by an emotion-arousing process that (a) generates a variation in their emotional state, and (b) triggers a behavioral and linguistic response. The concept of affective scenes has been proposed to explain the unfolding of basic emotions in conversations and applied to operator-customer call analysis. In Fig. 3 we show a state representation of the affective scene. Starting from an initial state (e.g. customer-operator greeting) one of the two speaker may manifest first his/her emotion (e.g. frustation) followed by transition into other states (e.g. anger). The conversation will end into either a *positive* or *negative* state.

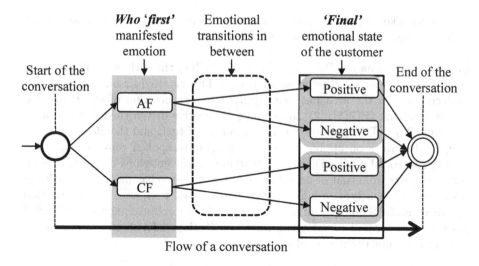

Fig. 3. State traces of affective scenes.

Affective scenes complement the linguistic scenes and their descriptions are being integrated to give a rich and complete description of the human conversations. An interesting extension of the two-party affective scene may be explored for multi-party conversations occurring in social media platforms.

Speech Overlaps. Another relevant topic we have investigated is *speech overlaps* and their semantic and discourse function. Speech overlaps are important events in spontaneous spoken conversations. In contact centers, speech overlap segments account for less than 10 % of the spoken segments [39] and they are required for *stitching* together the speech acts of speakers. Overlapping speech may reflect many aspects of discourse dynamics as well as emotional states. In [39, 40] we have focused on the pragmatic role of *competitive* or *non-competitive* overlaps and the roles of speakers in the act of overlapping. Further research will include the investigation of speech overlaps with respect to the semantic as well as affective description of human dialogues.

3.3   Discourse Structure and Coreference

It has long been known that the structure imposed on discourse by the relations underpinning its (relational) coherence is key to the human ability to recall and summarize information (e.g., [16,28]). This link was a key motivation for the early work on discourse structure and discourse parsing [13,14]. More recently, it has been shown that the information about entity coherence information [9,19] that can be extracted from text by intra-document and inter-document coreference resolution algorithms [20] also helps single-document and multi-document summarization [24,25] by identifying the main entities of a document or a collection of documents. These findings made the analysis of discourse structure and coreference a key aspect of SENSEI. In the following subsections we describe the main research topics we have addressed.

Domain Adaptation for Discourse Parsing. Much of the early work in discourse parsing was based on Rhetorical Structure Theory [13,14], but much of the modern work in the area has been spurred by the creation of the Penn Discourse Treebank (PDTB) [21], based on a connective-driven theory of discourse structure. The PDTB, however, consists mostly of text, and therefore there has been limited work on applying discourse parsing to spoken conversations, and even less to social media. Discourse structure in conversations differs in a number of respects from that of text. For instance, dialogue has more pragmatically motivated relation types when compared to written text, such as Interruption (speaker couldn't complete an utterance). Work on discourse parsing in SENSEI has therefore focused on adapting methods developed for discourse parsing to take into account the nature of speech [26,41].

Argument Structure. Among the relations found in conversations, those that specify the structure of arguments were expected to be of particular interest to SENSEI. Social media such as blogs or commentaries to newspaper articles have an inherently argumentative structure: people agree or disagree with a particular point being made. In order to properly understand such interactions it is essential to recognize which of the comments support the point of the commenter and which ones instead are opposed. Argumentation mining has gained increased interest in recent years [18,23,33]; much of this work has been applied to the classification of argumentative propositions in online user comments [1,4].

In SENSEI, we early on identified argument structure as an aspect of discourse structure of particular relevance to the task of summarizing online conversations, and have devoted substantial effort to it, by organizing a shared task on Online Forums Summarization at MULTILING-2015 that has focused on argument structure summarization [5,8] and by creating resources to support the task [2,8]. The shared-task annotation data may be obtained by contacting the consortium at [36].

Coreference. Intra-document coreference is the task of identifying the mentions that refer to the same entity within a document. Annotated corpora for this task became available in the mid-90s, enabling a great deal of research [20]. Recent corpora such as OntoNotes and ARRAU also moved away from annotation schemes motivated entirely by information extraction applications; systems trained on such corpora have been shown to work better for applications such as summarization that rely on some measure of text cohesion [25]. There has only been, however, limited work on coreference in spoken conversations and social media analysis, because of the lack of resources – to our knowledge, prior to SENSEI the LiveMemories-Blog corpus of Italian [22] was the only collection of social media data annotated for intra-document coreference, and we are aware of only one study of intra-document coreference for social media [12]. As in the case of discourse parsing, our primary objective was to develop methods to adapt models for coreference resolution trained on news to the conversation domain. We have carried out two lines of research in our work in this area. On the one hand, we have created annotated resources to study coreference in online forums, annotating for coreference the English and Italian datasets created for the Online Forums summarization task. On the other end, we have carried out research on domain adaptation for spoken conversations and social media data using our own BART platform, already tested in the 2011 and 2012 CONLL shared tasks [29,30]. Work so far includes adapting BART to run on French conversations [7] as well as work on domain adaptation for social media.

4 Summarization

There is a large body of work on text summarization, but very little that is specifically relevant to the analysis of human conversations in such diverse contexts as speech and social media. Good general overviews of automatic text summarization can be found in [59,60]. In the rest of this section we briefly review the related literature, discuss the novel research problems addressed by SENSEI and present preliminary results.

4.1 Speech Summarization

First approaches to spoken conversation summarization [32,35] have mostly focused on extractive summarization, which consists in selecting relevant utterances from the recordings and displaying their transcript to the user. Those approaches and all the extractive approaches proposed after them [3,6,10,34,59] have shown limitations in that they decontextualize the participants' discourse, and are unable to generalize and relate events discussed over a long time span.

In the call-center domain, in [61] they aim at automatically completing post-call logs, a type of summary generally manually created. The approach consists in filling templates with structured parts (detected from speech recordings) and unstructured parts created with extractive summarization methods. The authors show that call handling times are reduced without compromising log quality. [62] also address the problem of generating call-centre dialogue summaries, but with an unsupervised approach that performs topic induction and extracts utterances under an HMM model. Evaluation is only performed on synthetic dialogues. [63] adopt a different approach which leverages existing pairs of (speech recording, call log) through a method which associates utterances and log words. The method has a negative impact on call log quality, even though it outperforms other automatic baselines.

In SENSEI, we aim at going beyond extractive summarization in order to create abstractive descriptions of the content of conversations. In particular, abstractive summaries of call-centre conversations should be able to yield insight into why the customer called, what was her query, how did the agent solve that problem, was the behavior of the agent appropriate during the dialogue. We call such summaries *synopses*. They have two roles in the project: showing that we

Table 1. Example of synopses written by annotators for a single conversation from the Decoda French corpus [64] which includes calls from citizens enquiryng about public transportation.

Annotator	Synopsis
1	Request for itinerary from suburbs to downtown Paris. The caller wants to understand the fare given by one of his employees.
2	Request for information about the zones to take for a Navigo card for one person living in Chailly-en-Brie to travel in Paris. Zones 1 to 6.
3	An employer is calling the customer service cause he is not very sure about the ticket he has to pay for his employee. His employee is asking him for a sum which doesn't correspond to the fares and so he has the feeling that he is being ripped off.

have reached a sufficient understanding of the conversations, and creating a short textual representation of conversations that can be used to browse call-centre large databases, compare and group similar conversations, and help supervisors find conversations requiring more investigation. Examples of synopses are given in Table 1.

Unlike news summarization, which focuses on locating facts in text written by journalists and selecting the most relevant facts, conversation synopses require an extra level of analysis in order to achieve abstraction. Turn taking from the speakers has to be converted to generic expression of their needs, beliefs and actions. Even though extractive systems might give a glimpse of the dialogues, only abstraction can yield the story of what happens in the conversations.

Recent work on abstractive speech summarization includes modeling text generation as a Markov Decision Process [17] and generating a summary word by word, given a set of sentence clusters from the input. It is reminiscent of the recent trend towards conditioned language models [27,31] which use Recurrent Neural Networks for producing words. A similar approach [15] finds sentence communities through textual entailment and merges them. While those approaches are adequate when large quantities of annotated data are available, they are unsuitable for call-centre conversations which are focused and non-redundant.

Preliminary work on the project has yielded an approach for creating abstractive summaries from conversation transcripts. It uses domain knowledge to fill hand-written templates from entities detected in the conversation transcript using topic-dependent rules. For example, for the public transportation domain, we first cluster conversations by topic, and then write a template for each topic. Each template is a regular language with optional and repeatable parts. Slots are expressed as cross-template variables which need to be filled from the conversation (Table 2).

We performed evaluation on a subset of templates on the CCCS Shared Task for the Decoda corpus [64] using the ROUGE-2 evaluation metric [69]. The abstractive summarization systems are compared to extractive and

Table 2. Example of templates manually created for the Decoda French corpus (translated from French) [64]. We use the regular-expression formalism for denoting optional an repeatable parts.

Topic	Template
Itinerary	Query for itinerary (using $TRANSPORT)? from $FROM to $TO (without using $NOT_TRANSPORT)?. (Take the $LINE towards $TOWARDS from $START_STOP to $END_STOP.)*. Query for location $LOCATION.
Navigo pass	Query for (justification|refund|fares|receipt) for $CARD_TYPE. Customer has to go to offices at $ADDRESS.
Lost&found	$ITEM lost in $TRANSPORT (at $LOCATION)? (around $TIME)?. (Found, to be retrieved from $RETRIEVE_LOCATION |Not found).

abstractive baselines. The extractive baselines are the longest turn of the conversation, the longest turn in the first quarter of the conversation and Maximal Marginal Relevance (MMR). The first abstractive baseline consists of replacing the slot values with a bogus token which is not matched by Rouge during evaluation in order to simulate the worst slot filling system. The second baseline is based on the assumption that named entities play an important role in synopses: it consists in concatenating conversation named entities until the length constraint, without repetition. This baseline achieves a very bad readability, as expected. The topline consists in replacing the slot values with those manually annotated in the reference synopses. Results are summarized in Table 3.

In addition to hand-written templates, which fit well-structured conversations, we have addressed unexpected events through template generation. Following [65], additional templates are learned by extracting frequent patterns from hand-written synopses, generalizing slot variables and filling the templates with entities extracted from the conversation transcript. The generalization and template generation process includes *(a)* aligning synopses to conversation

Table 3. Rouge-2 results of the Decoda synopsis generation systems on a subset of the CCCS test set [64].

System	Rouge-2
Longest turn extract	0.04030
Longest turn @ 25 %	0.04594
MMR extract	0.04490
Hand-written templates + Bogus slots	0.02228
Named entities concatenation	0.09337
Hand-written templates + auto slots	0.10084
Abstractive topline	0.18067

sentences sharing the same semantic frames *(b)* mapping word tokens into their WordNet synsets and *(c)* clustering the generalized synopses to form the final templates.

4.2   Social Media Summarization

Previous work on summarization of text-based conversations and specifically of reader comment in on-line news is even more limited than that on summarization of spoken conversation. Summarization of email threads [54] and chat/on-line discussions [57] are similar tasks but there are critical differences. In the case of reader comment there is an initial news article that readers comment on and the relation of comments to this text is central – there is no direct analogue to this in the case of email or on-line discussion. Furthermore, email and on-line chat tend to involve longer exchanges between smaller numbers of correspondents in a more conventional dialogue form.

A small number of authors have directly addressed the task of summarizing on-line conversations commenting on videos or news articles. Khabiri et al. [50] addressed the task of summarising comments relating to Youtube videos. Ma et al. [51] addressed the task of summarising reader comments in on-line news, specifically *Yahoo! News* with a view to generating "an easy overview of all topics discussed in the comments". Llewellyn et al. [52] address the task of summarising reader comments in *The Guardian* newspaper and follow a similar approach to [50,51], again adopting a three stage process of topical clustering, ranking comments within clusters and then selecting top ranked comments across multiple clusters.

By contrast with earlier work that does not examine what form summaries of reader comments should take, in SENSEI we began by working with end users – journalists, news editors and readers and posters of reader comments – in a comprehensive study to identify use cases surrounding access to information in reader comments [48]. Six use cases were identified, including issue-oriented summaries of a single article+comment set, "blogger-oriented" summaries of all the postings of a single commenter and trend analysis summaries tracking issues across multiple article+comment sets over time.

We have chosen to focus initially on the use case of generating issue-oriented summaries of the comment set associated with a single article, a task bearing similarities to that of a journalist covering a town hall meeting. To support this work we have generated a set of gold standard human-authored summaries for a set of 18 article+comment sets, taking just the first 100 comments for each article [45]. This is the first set of such human-authored summaries for reader comments and the method and tools developed to create it as well as the resulting resource is a significant outcome of the project. Summary authors were given guidelines that, put briefly, instructed them to identify key issues discussed in the reader comments, positions taken with respect to these issues and the emotional tone of the discussion and to aggregate over these when writing their summaries. I.e., summaries were of the form "Many commenters discussed X with most taking stance S while a few took stance T. Other commenters debated Y in a very heated exchange with ...", capturing the issues discussed, the distribution of views on these issues and the affective character of the discussion. As a side of effect of the summary writing process, summary authors also grouped comments (around the issues discussed) with bi-directional links between comments, comment groups and summary sentences.

In SENSEI we have developed two approaches to automatically generating summaries of single article+comment sets. The first is an *extractive approach* that follows the same general line as previous work: clustering comments by topic, then ranking comments within clusters and finally selecting comments from within clusters to produce a final summary. However, there are several significant differences. First, we have developed a technique to link sentences in comments to sentences in the original article to which they are most similar, or none if the similarity is below a threshold [66]. This capability is used both in clustering (two comment sentences that link to the same article sentence are likely to be in the same cluster) and in summarization, where we have experimented with building summaries from comment clusters in different ways depending on whether or not the cluster contains any comments linked to the article (one might conjecture that summaries linked to the article are more on-topic/serious and hence more likely to contribute to issue-based summaries). Secondly, we have used a different method for clustering, the graph-based Markov Clustering Algorithm [67], leading to clustering results the significantly perform the state-of-the-art LDA-base approaches adopted to date. Finally, we have experimented with many different ranking methods.

Ranking and Extractive Summarization Results. Given a set of comment clusters, extractive summaries may be generated from them in a many different ways. Essentially this comes down to two separate ranking tasks: ranking clusters and ranking sentences within clusters. Summaries are then generated by visiting each cluster in ranked order and selecting from each the top-ranked sentence, until the summary length constraint is reached. We explored three classes of approach.

1. *Baseline Approaches:* No language processing is carried out. Threads are taken to be topically coherent comment groupings, so no clustering is used. Three variations of thread (cluster) sorting were considered: by time of first comment, by number of distinct participants and by number of comments. Comments within threads are sorted by time of posting. In this set of approaches sorting by number of comments worked best (**ParticipantCount-CentroidClosest**).

2. *Basic Text Processing Approaches:* Here again we take threads to be topically coherent comment groupings but consider 5 ways of ranking threads and 2 ways ranking comments within threads. Three ways of ranking threads are the same as used in the baseline approaches and in addition we consider ranking threads by cosine similarity of the thread centroid to the original news article (computed using a standard vector space model with each comment modelled as a vector) and by similarity of the thread centroid to the lead of the news article (first 5 sentences of the article). Within threads comments

Table 4. Summary evaluation results.

System	R1	R2	R-SU4
Human-Human	0.41	0.07	0.13
Time-CentroidClosest-Comment-in-Thread	0.35	0.04	0.10
ArticleLead-Sim-CentroidClosest-Comment-in-Thread	0.42	0.05	0.13
Linked-Cluster-ArticleLeadSim-Summary	0.40	0.04	0.12

are sorted either by time of posting or by cosine similarity of comments to thread centroid. Three of these 10 possible approaches are the same as the baseline approaches. Of the 7 new approaches the one that works best is ranking threads by similarity to the article lead and comments within a thread by similarity to the thread centroid (**ArticleLead-Sim-CentroidClosest-Comment-in-Thread**).

3. *Clustering and Article-Linking Approaches:* The final set of approaches make use of comment-article linking and comment clustering, as described above. A comment cluster is said to link to the original article if any of the comments in it link to the original article. This gives rise to three sets of clusters: linked clusters (all clusters are linked), unlinked clusters (no cluster is linked) and all clusters (linked or unlinked). We experimented with generating summaries from comments taken only from these different cluster sets and found best results were obtained by using just clusters from the linked set of clusters, sorting these cluster by cosine similarity of cluster centroid to article lead and then sorting sentences by cosine similarity to cluster centroid (**Linked-Cluster-ArticleLeadSim-Summary**).

To assess the quality of extractive summarization we use the gold standard summaries described above. We compared the automatically generated summaries against the model summaries using ROUGE [69] and using the standard measures of ROUGE 1 (R1), ROUGE 2 (R2) and ROUGE SU4 (RSU4). ROUGE 1 and 2 give recall scores for uni-gram and bi-gram overlap respectively between the automatically generated summaries and the reference ones. ROUGE SU4 allows bi-grams to be composed of non-contiguous words, with a maximum of four words between the bi-grams. The results of the summary evaluation are shown in Table 4 for the best of class system variants; full details may be found in [46].

The results show that one of the basic text processing approaches works best, one that does not bother with topical clustering but simply takes threads as topic clusters. Two caveats should be made, however. The first is that numerical differences here are small and may not be significant. The second is that as the gold standard summaries are abstractive summaries that feature aggregation over comments, ROUGE, which is fundamentally a lexical overlap measure, may not be appropriate as an intrinsic evaluation measure for this type of summary. The low human-human scores, as compared with the basic text processing approach, may support this sceptical view.

The second approach to summarization of reader comments associated with a single article is a *template-based approach*. Building on the definition of summary type for the issue-oriented or town hall summaries (see [46,48]), we defined a summary template consisting of the article title, a list of main topics discussed in the article and comments, the moods associated with the main topics, an indication of where opinion was consensual or divided, the most central topic and the key contributor to the discussion. The template is filled with data from three different modules: topic extraction, mood prediction and agreement/disagreement detection. Topic model is computed via the hierarchical Latent Dirichlet Allocation (LDA) over each news article and its user comments. The agreement/disagreement detection is based on the relation defined in the CorEA corpus of Italian reader comments *Corriere* [2]. Following the automatic topic linking, mood and agreement-disagreement relations prediction, a final template filling module writes out the template. Individual components on this approach have been evaluated in [46]. The running prototype can be viewed at [36].

On-going work on summarisation in SENSEI is now looking at moving beyond extractive and template-based approaches towards abstractive approaches that will take advantage of work on semantic parsing, parasemantic analysis and discourse and coreference analysis to generate summaries more akin to those that users have specified and that our gold standard exemplifies.

5   Evaluation of Summarization End-User Systems

In Sect. 4.2 above we have discussed the gold standard summary resource we created for evaluating reader comment summaries. This sort of resource is useful for *intrinsic* evaluation of summaries: it allows system developers to assess how close the summaries their systems produce are to what we believe a model summary to be. However, it does not tell us whether our summaries are helpful to end users in some task context. To do the latter we need to specify an *extrinsic* evaluation: a user task, a system or systems to assist the user with the task and metrics for assessing how well a user has performed at the task using the system(s). In SENSEI, the common approach to the extrinsic evaluation is to have the quality and usefulness of the summary to be assessed by the end-users. In the following sections we report on the evaluation frameworks for the speech and social media use cases.

5.1   Speech Use Case

For the evaluation of the SENSEI speech summarization prototype we follow an incremental evaluation model that includes the specification of the tasks, the selection and annotation of exemplar data and the comparative analysis of performances. The process is repeated over the development process of the prototype. Feedback from the evaluation cycles allows the assessment of the performance of the prototype, and the validation of the use cases.

In the speech scenario we have identified the Quality Assurance (QA) supervisors of a call centre as potential end-users. In contact centres the QA supervisors listen to the call and evaluate agents' compliance with the company protocol during the conversations with their customers. Agents' behaviour contributes to the overall quality of the calls, and the QA supervisors score the quality against established contact handling criteria, summarised into a QA monitoring form. In state-of-the-art business processes, the conversations are scored manually and results are recorded in the so-called Agent Conversation Observation Form (ACOF henceforth) [43]. This process may be both time consuming and sometimes inefficient due to the limited amount of calls that QA professionals can listen to every day. One of the goals of SENSEI is to automatically review and score operator-customer calls, and to summarise the features of the agents' behaviour in each call by an automatically generated QA form (e.g. the ACOF). Additionally, as discussed in Sect. 4.1, the goal is the automatic generation of short summaries (synopses) of each call. The speech use case evaluation has been carried over those two tasks.

For the ACOF generation task, the SENSEI prototype classifies the conversations on the basis of aspects of the agent's behaviour, such as the agent's ability to solve the customer problem, their empathic attitude, call resolution effectiveness, and so on. The goal is to evaluate the predictive performance of the SENSEI system in classifying the calls according to the ACOF criteria. We have designed an evaluation task where the automatic ratings assigned by the SENSEI prototype are compared with those assigned by human evaluators. In our case the human evaluators are QA analysts and supervisors. On average, evaluators find the SENSEI prototype is sufficiently accurate for the French and Italian corpus. The Likert ranking for both the Italian and the French corpus was 2.8. Details of this evaluation task can be found in [45].

For evaluating the SENSEI prototype with respect to the second task of synopsis generation, we have set up an extrinsic evaluation task. The task aims at identifying if, and to what extent, the availability of automatically generated summaries may help QA supervisors in mining conversation types such as problematic calls. Focusing on problematic calls is important because it may potentially reduce the time-to-completion of tasks related with the supervision of call centre agents. At present a great number of calls need to be listened to and assessed in order to identify the potentially problematic ones as soon as they occur in the call centre. The design of this task is based on a focus group methodology, whose goals are the discovery of shared views among the participants, and the implications behind those views for the SENSEI speech prototype. The evaluation task requires that the group participants should be representative of the potential population of users of SENSEI speech prototype. In [48] we identified quality assurance and human resources professionals as end-users, and participants form that user group has been recruited for the focus group.

In the Table 5 we report the end-user comments (right column) that have emerged from the discussions for each question (left column). In that discussion we had four participants plus the moderator: participants A and B were QA

Table 5. Comments on focus group questions. A, B are QA supervisors and C is a quality assurance manager.

Question	Comment
How was your experience while using ACOF ?	A and B reported a positive experience
ACOF could highlight agent's behaviour?	A and B gave a positive answer
Did you agree with the ratings of the automatically filled ACOFs?	Most of the time
Do you expect that SENSEI ACOFs may help you in saving time in your job?	A, B, and C agreed
Do you think ACOFs could be enriched with evidence of the system decisions?	A and B gave positive answers
Usefulness of the synopses of the call?	A, B, and C think synopses might be useful
Why synopses could be useful?	All: To assess first call resolution and reasons for inbound calls
What is SENSEI potential added value for your job?	All: SENSEI system may allow to supervise a larger number of calls

supervisors, participant C was a quality assurance manager, and participant D was an HR specialist. As for the turn taking within the focus group, the conversations have been smooth and the participants have been collaborative.

In general, the focus group participants have found that the SENSEI results could be useful for their job because they would allow a larger number of calls to be monitored. They have also recommended that the automatic selection of problematic calls could be useful for partially overcoming the biases of human evaluation.

5.2   Social Media Use Case

In the case of reader comment summarization, identifying a user task poses challenges. This is because no one currently writes summaries of these comments as part of some larger task nor is there an obvious current user task setting in which summaries of reader comments would prove helpful. That said, our user study [48] has revealed considerable interest in such summaries and a wide set of user types and task settings where such summaries might play a useful role. One user task that could prove useful across end-user types, is that of automatically generating an overview of the key issues discussed in a reader comment set and the positions taken on these issues. We have constructed the following task-based evaluation motivated by this scenario. Further details may be found in [44, 45]. To the best of our knowledge this is the first task-based evaluation protocol for reader comment summaries yet proposed.

Evaluation Tasks. We propose the following series of tasks for users to carry out in such an evaluation:

1. Overview Questions: first, we ask participants to play the role of a user wanting to make sense of a comment conversation in a short period of time, e.g. a coffee break; we then provide users with a system and a topic (an article and comment set); allow a set time for reading over news and comment (e.g. 2 min) and then ask users to: (1) identify four main issues in the discussion and (2) characterise opinion on a given issue in a set time (e.g. 10 min) in accordance with our definitions.
2. Post task questionnaire: we ask participants to rate and compare the usefulness of the system(s) and system components in the context of completing the tasks, on a five point scale and include an option for written feedback.
3. Finally, in a guided group discussion we invite participants to comment on their experience during the tasks and on using the different systems/components.

This protocol provides three complementary sets of results. To compare systems, we can now design experiments with any number of different system-variants, involving participants and topics as required, to control for topic effects and individual user differences. We then use the results of the protocol with each task instance to compare how, and to what extent, the different systems help users in carrying out the overview task.

A Pilot Evaluation. Participant responses to the overview questions are assessed manually. Assessors are given the source comments and the gold standard summaries (we select only articles which also appear in our gold standard for the extrinsic evaluation) and are asked to score written responses on a graded scale. The issues identified by participants in response to the overview questions are scored on a 4 point scale that takes account of criteria such as evidence/accuracy and clarity of expression. Characterisation of opinion is scored on a graded 6 point scale, based on criteria of coverage, representing quantities and accuracy. We analyze the free text and spoken responses gathered in the post task questionnaire and discussion using simple qualitative techniques. Data from the user ratings of the different systems/system components is summarised using simple statistics.

To carry out comparative evaluations of different systems we have developed a configurable interface with the following characteristics. It includes a baseline comment-only system, which presents threaded conversations in the way they typically appear in on-line news today, for example on *The Guardian* website. It takes as input comment clusters, labels for these clusters and summaries, which may be either extractive or abstractive and may contain links between sentences in the summary and the comment cluster that gave rise to the sentence. It offers two summary presentation modes: a text-based summary presentation mode and a graphical summary presentation mode. In the text-based mode the supplied summary and a textual representative of each cluster (e.g. a cluster label or representative phrase or sentence) are displayed. The sentences in the summary, if links to clusters are provided, are clickable allowing the clusters underlying the sentences to be displayed. The textual representative of the clusters are also clickable allowing the comments in the cluster to be displayed.

We have tested the full task protocol and interface in a pilot evaluation. Four participants, all post-graduates with experience in language technologies and using reader comment, each carried out two iterations of the task, each time using a different system/interface configuration:

S1. A baseline, presenting just the reader comment facility used by *The Guardian* in current practice.

S2. Included both the baseline functionality and sense-making components, consisting of a labelled pie chart indicating the relative size of comment clusters and a textual summary whose sentences were linked to underlying comment clusters. The clustering, cluster labelling and summarization outputs were produced by the top performing component combination described in Sect. 4.2 above, the ArticleLead-Sim-CentroidClosest-Comment-in-Thread system.

There were two different topics, each comprising a news article and an associated set of 100 comments. Each participant used each system and each topic exactly once. We provided a short training session including a system demo and guidelines on the overview scenario and tasks. We scored answers to the content questions using the metrics described above, aggregated ratings from the feedback questionnaire, and carried out a qualitative analysis of feedback from the group discussion. The three complementary sets of results allowed us to assess the protocol and to compare how, and to what extent, the different systems and system components helped users to complete the two content-related questions. While feedback on the usefulness of the sense-making technologies suggested more development was necessary if outputs were to help in such contexts, the general interface design and direction of the technology, as guided by the overview task, was approved of. The results also indicated that the protocol provides sufficient data to answer questions such as did different systems help with different content questions? Did one system help better overall? What features of the interface did users find most helpful in the task context? etc. A complete description of the methodology and evaluation task is given in [45].

6   Conclusion

The SENSEI project aims at taking a radically new approach at developing the technology for language summarization. We have selected a very relevant domain for the evaluation of the summarization technology: human conversations generated in contact centers and user comments on on-line news articles. By taking a *vertical* approach to the evaluation of the technology we have connected the end-users (e.g. customers or journalists) to the speech and natural language processing components and we expect to impact the efficacy of summary definition, generation and assessment. While improving the value of the summary on the end-user task, we have shown that new research on semantic, para-semantic and discourse parsing has greatly contributed to the automatic generation of a novel type of summaries.

Acknowledgments. The research leading to these results has received funding from the European Union - Seventh Framework Program (FP7/2007-2013) under grant agreement no 610916.

The research reported in this paper would have not been possible without the key contributions of the researchers, engineers and professionals at the University of Trento, Aix-Marseille University, University of Sheffield, University of Essex, Websays and Teleperformance.

References

1. Boltuzic, F., Snajder, J.: Back up your stance: recognizing arguments in online discussions. In: Proceedings of the 1st Workshop on Argumentation Mining, Baltimore, MD, pp. 49–58 (2014)
2. Celli, F., Riccardi, G., Ghosh, A.: Corea: Italian news corpus with emotions and agreement. In: Proceedings of CLIC, Pisa (2014)
3. Erol, B., Lee, D.S., Hull, J.: Multimodal summarization of meeting recordings. In: 2003 International Conference on Multimedia and Expo, ICME 2003, Proceedings, vol. 3, pp. III–25. IEEE (2003)
4. Ghosh, D., Muresan, S., Wacholder, N., Aakhus, M., Mitsui, M.: Analyzing argumentative discourse units in online interactions. In: Proceedings of the 1st Workshop on Argumentation Mining, Baltimore, MD, pp. 39–48 (2014)
5. Giannakopoulos, G., Kubina, J., Conroy, J.M., Steinberger, J., Favre, B., Kabadjov, M., Kruschwitz, U., Poesio, M.: Multiling 2015: multilingual summarization of single and multi-documents, on-line fora, and call-center conversations. In: Proceedings of SIGDIAL, Prague, pp. 270–274 (2015)
6. Gillick, D., Riedhammer, K., Favre, B., Hakkani-Tur, D.: A global optimization framework for meeting summarization. In: Proceedings of ICASSP 2009, Taiwan, pp. 4769–4772. IEEE (2009)
7. Kabadjov, M., Bechet, F., Favre, B., Kruschwitz, U., Poesio, M.: A coreference resolver for french spoken conversations (2015, submitted)
8. Kabadjov, M., Steinberger, J., Barker, E., Kruschwitz, U., Poesio, M.: Onforums: the shared task on online forum summarisation at multiling '15. In: Proceedings of FIRE (2015)
9. Kintsch, W., van Dijk, T.: Towards a model of discourse comprehension and production. Psychol. Rev. **85**, 363–394 (1978)
10. Lai, C., Renals, S.: Incorporating lexical and prosodic information at different levels for meeting summarization. In: Fifteenth Annual Conference of the International Speech Communication Association (2014)
11. Lin, C.Y.: Rouge: a package for automatic evaluation of summaries. In: Text Summarization Branches Out: Proceedings of the ACL-04 Workshop, pp. 74–81 (2004)
12. Lindmark, D.: Methods for lean, precision-oriented, and targeted coreference resolution. Ph.D. thesis, Uppsala University (2012)
13. Mann, W.C., Thompson, S.A.: Rhetorical structure theory: towards a functional theory of text organization. Text **8**(3), 243–281 (1988)

14. Marcu, D.: The Theory and Practice of Discourse Parsing and Summarization. MIT Press, Cambridge (2000)
15. Mehdad, Y., Carenini, G., Tompa, F.W., Ng, R.T.: Abstractive meeting summarization with entailment and fusion. In: Proceedings of the 14th European Workshop on Natural Language Generation, pp. 136–146 (2013)
16. Meyer, B.: The Organization of Prose and its Effects on Memory. North-Holland, Amsterdam (1975)
17. Murray, G.: Abstractive meeting summarization as a Markov decision process. In: Barbosa, D., Milios, E. (eds.) Canadian AI 2015. LNCS, vol. 9091, pp. 212–219. Springer, Heidelberg (2015)
18. Palau, R.M., Moens, M.F.: Argumentation mining. J. Artif. Intell. Law **19**(1), 1–22 (2011)
19. Poesio, M., Stevenson, R., Di Eugenio, B., Hitzeman, J.M.: Centering: a parametric theory and its instantiations. Comput. Linguist. **30**(3), 309–363 (2004)
20. Poesio, M., Stuckardt, R., Versley, Y.: Anaphora Resolution: Algorithms, Resources and Applications. Springer, Berlin (2016)
21. Prasad, R., Dinesh, N., Lee, A., Miltsakaki, E., Robaldo, L., Joshi, A., Webber, B.: The penn discourse treebank 2.0. In: Proceedings of LREC (2008)
22. Rodriguez, K.J., Delogu, F., Versley, Y., Stemle, E., Poesio, M.: Anaphoric annotation of wikipedia and blogs in the live memories corpus. In: Proceedings of LREC (poster) (2010)
23. Stab, C., Gurevych, I.: Annotating argument components and relations in persuasive essays. In: Proceedings of COLING, pp. 1501–1510 (2014)
24. Steinberger, J., Kabadjov, M., Poesio, M., Pouliquen, B., Steinberger, R.: WB-JRC-UniTN's participation in TAC 2009: update summarization and AESOP tracks. In: Proceedings of TAC, Washington, November 2009
25. Steinberger, J., Poesio, M., Kabadjov, M., Jezek, K.: Two uses of anaphora resolution in summarization. Inf. Process. Manage. **43**(6), 1663–1680 (2007). Special Issue on Summarization
26. Stepanov, E.A., Riccardi, G., Bayer, A.O.: The unitn discourse parser in conll 2015 shared task: token-level sequence labeling with argument-specific models. In: Proceedings of CONLL - Shared Task, Bejing, pp. 25–31 (2015)
27. Sutskever, I., Vinyals, O., Le, Q.V.: Sequence to sequence learning with neural networks. In: Advances in Neural Information Processing Systems, pp. 3104–3112 (2014)
28. Taboada, M., Mann, W.C.: Applications of rhetorical structure theory. Discourse Stud. **8**(4), 567–588 (2006)
29. Uryupina, O., Moschitti, A., Poesio, M.: Bart goes multilingual: the unitn/essex submission to the conll-2012 shared task. In: Proceedings of the 15th CONLL: Shared Task. Association for Computational Linguistics, Korea, July 2012
30. Versley, Y., Ponzetto, S., Poesio, M., Eidelman, V., Jern, A., Smith, J., Yang, X., Moschitti, A.: Bart: a modular toolkit for coreference resolution. In: Proceedings of ACL, Demo Session, Columbus, OH, June 2008
31. Vinyals, O., Toshev, A., Bengio, S., Erhan, D.: Show and tell: A neural image caption generator. arXiv preprint arXiv:1411.4555 (2014)
32. Waibel, A., Bett, M., Finke, M., Stiefelhagen, R.: Meeting browser: tracking and summarizing meetings. In: Proceedings of the DARPA Broadcast News Workshop, pp. 281–286. Citeseer (1998)
33. Walker, M., Tree, J.F., Anand, P., Abbott, R., King, J.: A corpus for research on deliberation and debate. In: Proceedings of LREC (2012)

34. Wang, L., Cardie, C.: Focused meeting summarization via unsupervised relation extraction. In: Proceedings of the 13th Annual Meeting of the Special Interest Group on Discourse and Dialogue, pp. 304–313. Association for Computational Linguistics (2012)
35. Zechner, K.: Automatic summarization of open-domain multiparty dialogues in diverse genres. Comput. Linguist. **28**(4), 447–485 (2002)
36. Making Sense of Human-Human Conversations Data, Project FP7/2007-2013. http://www.sensei-conversation.eu
37. Celikyilmaz, A., Hakkani-Tur, D., Pasupat, P., Sarikaya, R.: Enriching word embeddings using knowledge graph for semantic tagging in conversational dialog systems. In: AAAI - Association for the Advancement of Artificial Intelligence (2015)
38. Bordes, A., Usunier, N., Garcia-Duran, A., Weston, J., Yakhnenko, O.: Translating embeddings for modeling multi-relational data. In: Burges, C.J.C., Bottou, L., Welling, M., Ghahramani, Z., Weinberger, K.Q. (eds.) Advances in Neural Information Processing Systems 26, pp. 2787–2795 (2013)
39. Chowdhury, A., Danieli, M., Riccardi, G.: The role of speakers and context in classifying competition in overlapping speech. In: Proceedings of INTERSPEECH, Dresden (2015)
40. Chowdhury, A., Danieli, M., Riccardi, G.: Annotating and categorizing competition in overlap speech. In: Proceedings of ICASSP, Brisbane (2015)
41. Riccardi, G., Stepanov, E., Chowdhury, A.: Discourse connective detection in spoken conversations. In: Proceedings of ICASSP, Shangai (2016)
42. Danieli, M., Riccardi, G., Alam, F.: Emotion unfolding and affective scenes: a case study in spoken conversations. In: Proceedings of ICMI Workshop on Representations and Modeling for Companion Systems, Seattle (2015)
43. Danieli, M., Balamurali, A.R., Stepanov, E., Favre, B., Bechet, F., Riccardi, G.: Summarizing behaviours: an experiment on the annotation of call-centre conversations. In: Proceedings of LREC, Portoroz (2016, to appear)
44. Barker, E., Funk, A., Paramita, M., Kurtic, E., Aker, A., Foster, J., Hepple, M., Gaizauskas, R.: What's the issue here?: task-based evaluation of reader comment summarization systems. In: Proceedings of LREC, Portoroz (2016, to appear)
45. Danieli, M., Barker, E. (eds.): Report on Intermediate Evaluation, SENSEI Deliverable D1.3, October 2015
46. Aker, A. (ed.): Report on Specification of Conversation Analysis and Summarization Outputs, SENSEI Deliverable D5.2, October 2015
47. Bayer, A.O., Riccardi, G.: Deep semantic encoding for language modeling. In: Proceedings of INTERSPEECH, Dresden (2015)
48. Danieli, M., Gaizauskas, R. (eds.): Report on Use Case Design and User Requirements, SENSEI Deliverable D1.2, October 2014
49. Foster, I., Kesselman, C.: The Grid: Blueprint for a New Computing Infrastructure. Morgan Kaufmann, San Francisco (1999)
50. Khabiri, E., Caverlee, J., Hsu, C.-F.: Summarizing user-contributed comments. In: Proceedings of The Fifth International AAAI Conference on Weblogs and Social Media (ICWSM-2011), pp. 534–537 (2011)
51. Ma, Z., Sun, A., Yuan, Q., Cong, G.: Topic-driven reader comments summarization. In: Proceedings of the 21st ACM International Conference on Information and Knowledge Management (CIKM 2012), pp. 265–274 (2012)
52. Llewellyn, C., Grover, C., Oberlander, J.: Summarizing newspaper comments. In: Proceedings of the Eighth International AAAI Conference on Weblogs and Social Media, pp. 599–602 (2014)

53. Mikolov, T., Chen, K., Corrado, G., Dean, J.: Efficient Estimation of Word Representations in Vector Space, CoRR, 1301.3781 (2013)
54. Murray, G., Carenini, G.: Summarizing spoken and written conversations. In: Proceedings of the Conference on Empirical Methods in Natural Language Processing, pp. 773–782 (2008)
55. Sundermeyer, M., Ney, H., Schluter, R.: From feedforward to recurrent LSTM neural networks for language modeling. IEEE/ACM Trans. Audio Speech Lang. Process. **23**(3), 517–529 (2015)
56. Tafforeau, J., Artieres, T., Favre, B., Bechet, F.: Adapting lexical representation and OOV handling from written to spoken language with word embedding. In: Sixteenth Annual Conference of the International Speech Communication Association, INTERSPEECH (2015)
57. Uthus, D.C., Aha, D.W.: Plans toward automated chat summarization. In: Proceedings of the Workshop on Automatic Summarization for Different Genres, Media, and Languages, pp. 1–7 (2011)
58. Vukotic, V., Raymond, C., Gravier, G.: Is it time to switch to word embedding and recurrent neural networks for spoken language understanding? In: InterSpeech (2015)
59. Nenkova, A., McKeown, K.: Automatic summarization. Found. Trends Inf. Retrieval **5**(2–3), 103–233 (2011)
60. Carenini, G., Murray, G., Ng, R.: Methods for Mining and Summarizing Text Conversations. Morgan and Claypool Publishers, San Rafael (2011)
61. Byrd, R.J., Neff, M.S., Teiken, W., Park, Y., Cheng, K.S.F., Gates, S.C., Visweswariah, K.: Semi-automated logging of contact center telephone calls. In: Proceedings of CIKM (2008)
62. Higashinaka, R., Minami, Y., Nishikawa, H., Dohsaka, K., Meguro, T., Takahashi, S., Kikui, G.: Learning to model domain-specific utterance sequences for extractive summarization of contact center dialogues. In: Proceedings of COLING (2010)
63. Tamura, A., Ishikawa, K., Saikou, M., Tsuchida, M.: Extractive summarization method for contact center dialogues based on call logs. In: Proceedings of IJCNLP (2011)
64. Favre, B., Stepanov, E., Trione, J., Bechet, F., Riccardi, G.: Call centre conversation summarization: a pilot task at multiling. In: Sigdial (2015)
65. Oya, T., Mehdad, Y., Carenini, G., Ng, R.: A template-based abstractive meeting summarization: leveraging summary and source text relationships. In: Proceedings of International Conference on Natural Language Generation (INLG 2014) (2014)
66. Aker, A., Kurtic, E., Hepple, M., Gaizauskas, R., Di Fabbrizio, G.: Comment-to-article linking in the online news domain. In: Proceedings of the 16th Annual Meeting of the Special Interest Group on Discourse and Dialogue (SIGDIAL 2015), pp. 245–249 (2015)
67. Aker, A., Kurtic, E., Balamurali, A.R., Paramita, M., Barker, E., Hepple, M., Gaizauskas, R.: A graph-based approach to topic clustering for online comments to news. In: Ferro, N., Crestani, F., Moens, M.-F., Mothe, J., Silvestri, F., Di Nunzio, G.M., Hauff, C., Silvello, G. (eds.) ECIR 2016. LNCS, vol. 9626, pp. 15–29. Springer, Switzerland (2016)
68. Das, M.K., Bansal, T., Bhattacharyya, C.: Going beyond Corr-LDA for detecting specific comments on news & blogs. In: Proceedings of the 7th ACM International Conference on Web Search and Data Mining, pp. 483–492 (2014)
69. Lin, C.-Y.: Rouge: a package for automatic evaluation of summaries. In: Proceedings of the ACL 2004 Workshop on Text Summarization Branches Out (2004)

Speech and Dialogue Technologies, Assets for the Multilingual Digital Single Market

Pierre-Paul Sondag[✉]

Programme Officer at the European Commission, DG CONNECT,
Luxembourg, Luxembourg
Pierre-Paul.Sondag@ec.europa.eu

1 The Early Days

In the early days, Natural Language Processing, signal processing and Artificial Intelligence were developed as separate strands of technologies, with few interactions between them. Researchers in these respective communities had different backgrounds; translation and linguistics for those in Natural Language Processing (NLP), statistics and engineering for those in signal processing, and finally cognitive sciences for the Artificial Intelligence ones. These R&D communities pursued their own goals, solving problems centred on a challenge specific to one community that didn't require much input from the others, for instance achieving textual machine translation, which could be carried out by language technology developers having only marginally to require contribution from other domains of expertise. Similarly R&D in signal processing focussed on single functions like speech recognition, speech generation, image recognition or virtual characters generation without requiring much contribution from NLP or AI.

1.1 Combining Technologies to Achieve Human-Machine Interaction

Over time, technologies from these different strands were combined to widen the abilities of computers to interact with humans and enhance their performance. Thus the interaction could evolve from a pre-established succession of well predictable action-result pairs between user and machine, towards less "hard-coded" interactions that were more spontaneous and came closer to our human way of interacting as we know it when we talk to other persons.

Natural dialogue between people is fuzzy and redundant, it implies vagueness and ambiguity, requires repetition, redundancies and reformulation. Natural dialogue works through successive deepening interaction loops to reach mutual understanding. Combining Language Technologies, speech processing and semantics/Artificial Intelligence into complex systems gives to the machine the ability to simulate our human behaviours.

Today human-machine or computer mediated human to human dialogue systems allow more natural and spontaneous ways of interacting. Thus the dialogue module became the glue that brought together and intertwined the different

© Springer International Publishing Switzerland 2016
J.F. Quesada et al. (Eds.): FETLT 2015, LNAI 9577, pp. 34–38, 2016.
DOI: 10.1007/978-3-319-33500-1_3

strands of technologies, increasing interaction performance and usability up to a level appropriate for their acceptance by users and the needs of real world applications.

1.2 Making Speech Interaction Multi-lingual

A noticeable attempt to make interaction multilingual was made by a large project funded by the German national scheme called Verbmobil[1] (1993–2000). It achieved speech to speech simultaneous interpretation between German and Japanese, based on the use case of a traveller making travel arrangements over the telephone with a person speaking in the other language. This project combined speech technologies with NLP technologies, hence symbolic processing and machine learning, achieving synergies out of this combination. Moreover the quality of interaction could be improved by having the system aware of the context.

Massive data collections were used to train the system, which put to the fore that speech-to-speech translation systems for spontaneous dialogs depends critically on the quantity and quality of the training corpora.

1.3 Making Interaction Multi-modal

Speech remains the most common and natural way to interact. However humans when talking to each other, exchange not only the words uttered by their mouth, they use many other clues to transmit their message to each other like face mimics, gaze, body language or by modulating the tone of their voice. Communication restricted solely to words remained a strong limitation of the initial speech interaction systems. Incidentally, they built among the general public the image of contrived artificial speech systems, lacking the most elementary behaviour expected by users, generating frustration and a reluctance to adopt these initial speech interaction systems.

To improve the usability, interaction systems were enhanced with other modalities. Among others, the Companion[2] Integrated Project (2006–2010) aimed to explore possible changes in the relationships people have with computers. The project developed two Embodied Conversational Agents (ECA) providing some sort of social interaction beyond the traditional task-oriented interaction. The English demonstrator *"How Was Your Day"* engaged a social conversation in which the user talked to the computer about his/her day at the office while the Czech *"Photopal"* engaged a social conversation with seniors using old photos albums as support for this social conversation.

Companion tested the inclusion of emotions and other modalities beyond speech in the interaction. The users' emotional state was detected based on

---

[1] Verbmobil was a long-term project of the German Federal Ministry of Education, Science, Research and Technology (BMBF, Projektträger DLR) http://verbmobil. dfki.de/.

[2] Integrated Project n° 610990 of the EU Sixth Framework Programme, http://cordis. europa.eu/project/rcn/110628_en.html.

his/her voice tone and prosody. While not yet directly applicable in the real world, the project illustrated the potential of including multimodality and affective computing in the cognitive strategy of the dialogue. It brought the ECA interaction a step closer to naturalness.

2   Improving Usability

2.1   Shift to a Data Driven Approaches

In the last decade technologies underlying conversational interaction systems became data driven. For instance speech that used to be synthesised by rule–based mapping of phonemes/triphones with syllables, is now synthesised in the Simple4All[3] project (2011–2014), by the means of slightly supervised machine learning techniques based on large corpora of speech resources. This makes the synthesised speech more expressive and more natural, while it also facilitates the personalization, enlarging the choice of voices and enabling to adapt them to a specific use-case. Moreover, the data driven approach requires far less human work in the development of the speech synthesizer and hence makes speech synthesis commercially viable even for languages corresponding to smaller markets. This appears particularly important for Europe in order to maintain our language diversity, as opposed to the past, where Language Technologies used to cover adequately only a small set of the spoken languages, corresponding to the ones spoken in the largest countries.

2.2   Understanding the Dialogue's Context

Data driven approaches are now also applied to the very core of conversational interaction systems (CIT), which is the dialogue management part. The Parlance[4] project (2011–2014) implemented a data driven dialogue management for conversational search systems that dynamically adapts to the users and the context. A prototype was built as a tourist information service dialoguing with the user in order to refine the user query until the system could provide a relevant result. The user triggered a session through a general question, for instance asking the system for a suggestion of a restaurant, after that the dialogue module interacted with him/her in order to refine the question in terms of location, genre and food type of the requested facility. The data driven approach enabled the system to dynamically adapt to the context including when new unpredicted situations arose, and to personalize the interaction to the user needs.

---

[3] Project n° 287678 of the EU Seventh Framework Programme, Speech synthesis that improves through adaptive learning, coordinated by the University of Edinburgh, http://simple4all.org/.

[4] Project n° 287615 of the EU Seventh Framework Programme, Probabilistic Adaptive Real-Time Learning And Natural Conversational Engine, coordinated by Heriot-Watt University, https://sites.google.com/site/parlanceprojectofficial.

2.3 Understanding the User's Intentions

An on-going project, Metalogue[5], explores further the interaction by adapting its own behaviour and trying to understand the user's behaviour, thus the focus is on providing meta-cognitive abilities to the dialogue system. One of the use cases used to validate the approach is a presentation trainer dedicated to train people for public speaking. The presentation trainer interacts with the user beyond the mere meaning of the words uttered by both user and system, it analyses also the user's voice (prosody, hums ...) and body language (facial expressions, posture, gestures ...) in order to evaluate his/her behaviour and give him/her advice for improvements.

Another use case is a meta-cognitive training app for smart-phone or tablet based on a game. The app develops the cognitive awareness of tactics of the human player by helping him to understand and anticipate the strategies of the opponent player.

The ability of conversational interaction systems to anticipate the behaviour of their human dialogue partner, hence to pro-actively support the interaction, is an essential precondition to make them perceived to be natural and spontaneous.

2.4 Making Conversational Systems to Behave More Like Humans

There is still a long way to go before conversational systems behave in a way really close to humans. As explored by the projects listed above, conversational systems need to move from a rigid task oriented dialogue structure to context and user behaviour aware dialogue that is dynamically generated. They have to perform simultaneously task related actions and communicative ones, like we as humans, do. They have to take advantage of all modalities, including emotions, body language, facial expression, gaze and finally to tackle metacognitive capabilities. Only the combination of many of these features will lead to a wider acceptance and use by the general public.

2.5 Planning the Way Forward

The European Commission through DG CONNECT launched several road mapping actions to plan the way forward. The ROCKIT support action developed a vision on research and innovation in the area of natural conversational interaction. It was supported by an on-line interactive roadmap[6] and networked a community of players active in that field. Two further support actions, CRACKER and LT-Observatory, developed together a Strategic Agenda for the multilingual Digital

---

[5] Project n° 611073 of the EU Seventh Framework Programme, Multiperspective Multimodal Dialogue, coordinated by DFKI Saarbrücken, http://www.metalogue.eu/.

[6] Project n° 611092 of the EU Seventh Framework Programme, Roadmap for Conversational Interaction Technologies, coordinated by the University of Edinburgh, 2014–2015 http://www.lt-innovate.org/citia.

Single Market[7] that combined the contribution and prospect of Language Technologies (LT) from respectively research and industry/innovation points of view.

3   Key Enablers of the Multilingual Digital Single Market

Applied together with localisation and machine translation, Conversational Interaction Technologies (CIT) will provide for all people, including to the less computer literate, access to digital services, and hence ease the advent of the multilingual Digital Single Market. The latter will be a means in the future to create economic growth and jobs, and will allow for more personalised, hence better products and services. Let's stress also the ability to preserve European values by providing people access to e-services in their own native language.

3.1   Parallels Between Conversational Interaction and Big Data

Parallels can be drawn between the development in conversational interaction technologies and the overall trend towards a data driven economy, as reflected in discussions about Big Data. The conversational interaction bridges different technologies in a similar way Big Data breaks the silos of application domains. Fusion of modalities in CIT echoes the data fusion in Big Data, while both need to overcome a similar challenge, the scalability. CIT are faced with the scalability challenge of the domain knowledge where Big Data faces the challenge of scalability of data integration. These parallels suggest that the domain of conversational technologies will be gradually extended to data-driven speech analytics, and that synergies could be found in future. CIT could thus benefit from the great interest aroused by Big Data and by the related resources made available.

3.2   EU Funding Opportunities in the Horizon 2020 Framework Programme

The current EU Framework Programme Horizon 2020 has currently no topics specifically dedicated to Language Technologies or CIT; nevertheless opportunities exist in several Societal Challenges topics and the introduction recommends their use in actions submitted for funding under the topics related to Big Data (ICT-14, ICT-15 and ICT-16)[8]. Proposers are encouraged to make use of Language Technologies, including machine translation, speech recognition, dialogue management, if the proposal involves analysis of information expressed in human language, or if the proposal addresses human-to-human or human-to-machine interaction or communication.

Thus some funding opportunities for CIT and LT exist in the Horizon 2020 programme; they could become more central in future as the CIT will be instrumental in the framework of larger data driven applications.

---

[7] Strategic Agenda for the multilingual Digital Single Market: http://cracker-project. eu.

[8] Introduction of the part 5i Information and Communication Technologies of the Horizon2020 Work Programme 2016 – 2017.

Regular Papers

Helping Domain Experts Build Phrasal Speech Translation Systems

Manny Rayner[1]([✉]), Alejandro Armando[1], Pierrette Bouillon[1], Sarah Ebling[2], Johanna Gerlach[1], Sonia Halimi[1], Irene Strasly[1], and Nikos Tsourakis[1]

[1] FTI/TIM, University of Geneva, Geneva, Switzerland
Emmanuel.Rayner@unige.ch
[2] Institute of Computational Linguistics, University of Zurich, Zurich, Switzerland
http://www.unige.ch/traduction-interpretation/
http://www.cl.uzh.ch/

Abstract. We present a new platform, "Regulus Lite", which supports rapid development and web deployment of several types of phrasal speech translation systems using a minimal formalism. A distinguishing feature is that most development work can be performed directly by domain experts. We motivate the need for platforms of this type and discuss three specific cases: medical speech translation, speech-to-sign-language translation and voice questionnaires. We briefly describe initial experiences in developing practical systems.

Keywords: Speech translation · Medical translation · Sign language translation · Questionnaires · Web

1 Introduction and Motivation

In this paper, we claim that there is a place for limited-domain rule-based speech translation systems which are more expressive than fixed-phrase but less expressive than general syntax-based transfer or interlingua architectures. We want it to be possible to construct these systems using a formalism that permits a domain expert to do most of the work and immediately deploy the result over the web. To this end, we describe a new platform, "Regulus Lite", which can be used to develop several different types of spoken language translation application.

A question immediately arises: are such platforms still relevant, given the existence of Google Translate (GT) and similar engines? We argue the answer is yes, with the clearest evidence perhaps coming from medical speech translation. Recent studies show, unsurprisingly, that GT is starting to be used in hospitals, for the obvious reason that it is vastly cheaper than paid human interpreters [4]; on the other hand, experience shows that GT, which has not been trained for this domain, is seriously unreliable on medical language. A recent paper [23] describes

Medical translation work was supported by Geneva University's Innogap program. Work on sign language translation was supported by the Crédit Suisse, Raiffeisen, TeamCO and Max Bircher Foundations. We thank Nuance Inc. and the University of East Anglia for generously allowing us to use their software for research purposes.

© Springer International Publishing Switzerland 2016
J.F. Quesada et al. (Eds.): FETLT 2015, LNAI 9577, pp. 41–52, 2016.
DOI: 10.1007/978-3-319-33500-1_4

the result of a semi-formal evaluation, in which it was used to translate ten text sentences that a doctor might plausibly say to a patient into 26 target languages. The bottom-line conclusion was that the results were incorrect more than one time in three.

Doctors are thus with good reason suspicious about the use of broad-coverage speech translation systems in medical contexts, and the existence of systems like MediBabble[1] gives further grounds to believe that there is a real problem to solve here. MediBabble builds on extremely unsophisticated translation technology (fixed-phrase, no speech input), but has achieved considerable popularity with medical practitioners. In safety-critical domains like medicine, there certainly seem to be many users who prefer a reliable, unsophisticated system to an unreliable, sophisticated one. MediBabble is a highly regarded app because the content is well-chosen and the translations are known to be good, and the rest is viewed as less important. The app has been constructed by doctors; a language technologist's reaction is that even if GT may be too unreliable for use in hospitals, one can hope that it is not necessary to go back to architectures quite as basic as this. A reasonable ambition is to search for a compromise which retains the desirable property of producing only reliable output prechecked by professional translators, but at the same time supports at least some kind of productive use of language, and also speech recognition.

A second type of application which has helped motivate the development of our architecture is speech-to-sign-language translation. Sign languages are low-resource, a problem they share with many of the target languages interesting in the context of medical speech translation. In addition, since they are non-linear, inherently relying on multiple parallel channels of communication including hand movement, eye gaze, head tilt and eyebrow inflection [21], it is not possible to formalise translation as the problem of converting a source-language string into a target-language string. It is in principle feasible to extend the SMT paradigm to cover this type of scenario, but currently available mainstream SMT engines do not do so. As a result, most previous SMT approaches to sign language machine translation, such as [19,27], have used unilinear representations of the sign languages involved. If we want to build sign-language translators which can produce high-quality output in the short-term, rule-based systems are a logical choice.

A third application area where this kind of approach seems appropriate is interactive multilingual questionnaires. Particularly in crisis areas, it is often useful for personnel in the field to be able to carry out quick surveys where information is elicited from subjects who have no language in common with the interviewer [26]. Again, applications of this kind only need simple and rigid coverage, but accurate translation and rapid deployability are essential, and practically interesting target languages are often underresourced.

In the rest of the paper, we describe Regulus Lite, showing how it can be used as an economical tool for building spoken language translation applications at least for the three domains we have just mentioned. The main focus is application content development. Section 2 gives an overview of the platform and the

---

[1] http://medibabble.com/.

rule formalism. Section 3 presents specific details on medical speech translation, sign language translation and voice questionnaires, and briefly sketches the initial applications. Section 4 presents some initial evaluation results for the voice questionnaire app, currently the most advanced one. The final section concludes.

2 The Platform

The Regulus Lite platform supports rapid development and web deployment for three types of small to medium vocabulary speech translation applications: plain translation, sign language translation, and voice questionnaires. We briefly describe each of these:

Plain Translation. The simplest case: the source language user speaks and the system displays its understanding of what the user said (a paraphrase of what was recognised). If the source language user approves the paraphrase, a target language translation is produced.

Sign Language Translation. Similar to plain translation, but the output is rendered in some form of sign language, using a signing avatar.

Voice Questionnaires. The content is organized as a form-filling questionnaire, where the interviewer poses the questions in spoken form, after which they are translated into the target language and presented to the subject. There are typically many possible questions for each field in the questionnaire. The subject responds by pressing one of a question-dependent set of buttons, each of which is labelled with a possible answer.

A basic assumption is that the content will be in the form of flat phrasal regular expression grammars. Reflecting this, content is specified using two basic constructions, TrPhrase (phrases) and TrLex (lexical items). Each construction combines one or more Source language patterns and at most one Target language result for each relevant target language, and indicates that the Source line can be translated as the Target. A trivial example[2] might be

```
TrPhrase $$top
Source ( hello | hi )
Target/french Bonjour
EndTrPhrase
```

A slightly more complex example, which includes a TrLex, might be

```
TrPhrase $$top
Source i ( want | would like ) $$food-or-drink ?please
Source ( could | can ) i have  $$food-or-drink ?please
Target/french je voudrais $$food-or-drink s'il vous plaît
EndTrPhrase

TrLex $$food-or-drink source="a (coca-cola | coke)" french="un coca"
```

---

[2] The notation has been changed slightly for expositional purposes.

Here, the variable `$$food-or-drink` in the first rule indicates a phrase that is to be translated using the second rule.

In order to decouple the source language and target language development tasks, TrPhrase and TrLex units are split into pieces placed in separate language-specific files, one for the source language and one for each target language. The connecting link is provided by a canonical version of the source language text (the portions marked as `Target/english` or `english=`). Thus the TrPhrase and TrLex units above will be reconstituted from the source-language (English) pieces

```
TrPhrase $$top
Source i ( want | would like ) $$food-or-drink ?please
Source ( could | can ) i have  $$food-or-drink ?please
Target/english i want $$food-or-drink please
EndTrPhrase
```

```
TrLex $$food-or-drink source="a (coca-cola | coke)" english="a coke"
```

and the target language (French) pieces

```
TrPhrase $$top
Target/english i want $$food-or-drink please
Target/french je voudrais $$food-or-drink s'il vous plaît
EndTrPhrase
```

```
TrLex $$food-or-drink english="a coke" french="un coca"
```

The development process starts with the source language developer writing their piece of each unit, defining the application's coverage. A script then generates "blank" versions of the target language files, in which the canonical source lines are filled in and the target language lines are left empty; so the French target language developer will receive a file containing items like the following, where their task is to replace the question marks by translating the canonical English sentences.

```
TrPhrase $$top
Target/english i want $$food-or-drink please
Target/french ?
EndTrPhrase
```

```
TrLex $$food-or-drink source="a coke" french="?"
```

As the source language developer adds more coverage, the "blank" target language files are periodically updated to include relevant new items.

The content can at any time be compiled into various pieces of runtime software, of which the most important are an application-specific grammar-based speech recogniser and a translation grammar; the underlying speech recognition engine used in the implemented version of the platform is Nuance Recognizer version 10.2. These generated software modules can be immediately uploaded to a webserver, so that the system is redeployable on a time scale of a few minutes. Applications can be hosted on mobile platforms — smartphones, tablets or laptops — linked over a 3G connection to a remote server, with recognition performed on the server [12]. The deployment-level architecture of the platform is adapted from that of the related platform described in [25], and offers essentially the same functionality.

3 Types of Application

3.1 Medical Translation

As already mentioned, medical speech translation is one of the areas which most strongly motivates our architecture. Several studies, including earlier projects of our own [23,30], suggest that doctors are dubious about the unpredictability of broad-coverage SMT systems and place high value on translations which have been previously validated by professional translators. Other relevant factors are that medical diagnosis dialogues are sterotypical and highly structured, and that the languages which pose practical difficulties are ones badly served by mainstream translation systems.

The practical problems arise from the fact that the Lite formalism only supports regular expression translation grammars. The question is thus what constituents we can find which it is safe always to translate compositionally. It is clear that many constituents cannot be treated in this way. Nonetheless, it turns out that enough of them can be translated compositionally that the grammar description is vastly more efficient than a completely enumerative framework; most adjuncts, in particular PPs and subordinate clauses, can be regarded as compositional, and it is often possible to treat nouns and adjectives compositionally in specific contexts.

We are currently developing a prototype medical speech translator in a collaboration with a group at Geneva's largest hospital[3]. Initial coverage is organised around medical examinations involving abdominal pain, with the rules loosely based on those developed under an earlier project [2]. Translation is from French to Spanish, Italian and Arabic[4]. A typical source language rule (slightly simplified for presentational purposes) is

```
TrPhrase $$top
Source ?$$PP_time la douleur est-elle ?$$adv $$qual ?$$PP_time
Source ?$$PP_time avez-vous ?$$adv une douleur $$qual
Source ?$$PP_time ?(est-ce que) la douleur est ?$$adv $$qual ?$$PP_time
Target/french la douleur est-elle ?$$adv $$qual ?$$PP_time
EndTrPhrase
```

Here, the French Source lines give different variants of *la douleur est-elle $$qual* ("Is the pain $$qual?"), for various substitutions of the transfer variable $$qual (*vive*, "sharp"; *difficile à situer*, "hard to localize"; *dans l'angle costal*, "in the intercostal angle", etc.). Each variant can optionally be modified by an adverb

---

[3] Hôpitaux Universitaires de Genéve.
[4] Tigrinya will be added soon.

($$adv) and/or a temporal PP ($$PP_time). Thus the questions covered will be things like *avez-vous souvent une douleur vive le matin?* ("do you often experience a sharp pain in the morning?") As the rule illustrates, there are typically many possible ways of formulating the question, all of which map onto a single canonical version. The target language translators work directly from the canonical version.

The current prototype represents the result of about one person-month of effort, nearly all of which was spent on developing the source side rules. Coverage consists of about 250 canonical patterns, expanding to about 3M possible source side sentences; the source language vocabulary is about 650 words. Creating a set of target language rules only involves translating the canonical patterns, and is very quick; for example, the rules for Italian, which were added at a late stage, took a few hours.

Speech recognition is anecdotally quite good: sentences which are within coverage are usually recognised, and correctly recognised utterances are always translated correctly. The informal opinion of the medical staff who have taken part in the experiments is that the system is already close to the point where it would be useful in real hospital situations, and clearly outperforms Google Translate within its intended area of application. We are in process of organising a first formal evaluation and expect to be able to report results in 2016.

3.2   Sign Language Translation

The rapidly emerging field of automatic sign language translation poses multiple challenges [3,5,7,9,13,16–18,20,22,28]. An immediate problem is that sign languages are very resource-poor. Even for the largest and best-understood sign languages, ASL and Auslan, the difficulty and expense of signed language annotation means there is an acute shortage of available corpus data[5]; for most of the world's estimated 120 sign languages [31], there are no corpora at all. In addition, there are often no reliable lexica or grammars and no native speakers of the language with training in linguistics.

Sign languages also pose unique challenges not shared with spoken languages. As already mentioned, they are inherently non-linear; even though the most important component of meaning is conveyed by the hands/arms (the *manual activity*), movements of the shoulders, head, and face (the *non-manual components*) are also extremely important and are capable of assuming functions at all linguistic levels [6]. Commonly cited examples include the use of head shakes/eyebrow movements to indicate negation and eye gaze/head tilt to convey topicalization [14,21]. Anecdotally, signers can to some extent understand signed language which only uses hand movements, but it is regarded as unnatural and can easily degenerate into incomprehensibility [29]; quantitatively, controlled studies show that the absence of non-manual information in synthesized signing

---

[5] The largest parallel corpus used in sign language translation that we know of has about 8 700 utterances [11].

(sign language animation) leads to lower comprehension scores and lower subjective ratings of the animations [15]. In summary, it is unsatisfactory to model sign language translation with the approximation most often used in practice: represent a signed utterance as a sequence of "glosses" (identifiers corresponding to hand signs), and consider the translation problem as that of finding a sequence of glosses corresponding to the source language utterance [8]. This approximation is unfortunately necessary if mainstream SMT engines are to be used.

For the above reasons and others, it is natural to argue that current technology requires high-quality automatic sign language translation to use rule-based methods in which signed utterances are represented in nonlinear form [13]. Our treatment conforms to these intuitions and adapts them to the minimalistic Lite framework. Following standard practice in the sign language linguistics literature, a signed utterance is represented at the linguistic level as a set of aligned lists, one for each parallel output stream: at the moment, we use six lists respectively called gloss (hand signs), head (head movements like nodding or shaking), gaze (direction of eye gaze), eyebrows (raising or furrowing of eyebrows), aperture (widening or narrowing of eyes) and mouthing (forming of sound-like shapes with the mouth).

gloss	TRAIN	CE	GENEVE	ALLER	PAS
head	Down	Down	Neutral	Neutral	Shaking
gaze	Neutral	Down	Neutral	Neutral	Neutral
eyebrows	FurrowBoth	FurrowBoth	Up	Up	Neutral
aperture	Small	Small	Neutral	Wide	Neutral
mouthing	Tr@	SS	Genève	Vais	Pas

Fig. 1. Sign table representation of an utterance in Swiss French Sign Language meaning "This train does not go through Geneva".

The examples we show below are taken from our initial application, which translates train service announcements from spoken French to Swiss French sign language. A typical sign table is shown in Fig. 1; translation from speech to sign is performed in three stages, with sign tables like these acting as an intermediate representation or pivot. As before, the first stage is to use speech recognition to produce a source language text string[6]. In the second, the source language string is translated into a sign table. Finally, the sign table is translated into a representation in SiGML [9], which can be fed into a signing avatar; in the current version of the system, we use JASigning [10]. The image below shows the user interface. On the left, we have, from top to bottom, the recognition result and the sign table; on the right, the avatar, the avatar controls and the SiGML.

---

[6] This is a slight oversimplification; in actual fact, recognition passes an n-best hypothesis list. The complications this introduces are irrelevant in the present context.

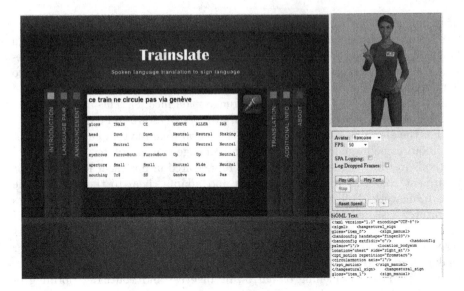

The issues that are of interest here are concerned with the text to sign table translation stage; once again, the central challenge is to create a formalism which can be used by linguists who are familiar with the conventions of sign language linguistics, but not necessarily with computational concepts. The formalism used is a natural generalization of the one employed for normal text-to-text translation; the difference is that the output is not one list of tokens, but six aligned lists, one for each of the output streams. For practical reasons, since correct alignment of the streams is crucial, it is convenient to write rules in spreadsheets and use the spreadsheet columns to enforce the alignment.

The non-obvious aspects arise from the fact that phrasal sign translation rules in general fail to specify values for all the output streams, with the values of the other streams being filled in by phrases higher up in the parse tree. Figure 2 illustrates, continuing the example from the previous figure. The lexical entry for *genéve* only specifies values for the **gloss** and **mouthing** lines. When the rules are combined to form the output shown in Fig. 1, the value of **eyebrows** associated with the sign glossed **GENEVE** is inherited from the phrase above, and thus becomes **Up**.

The process by which sign tables are translated into SiGML is tangential to the main focus of this paper, so we content ourselves with a brief summary. The information required to perform the translation is supplied by three lexicon spreadsheets, maintained by the sign language expert, which associate glosses and other identifiers appearing in the sign table with SiGML tags and strings written in HamNoSys [24], a popular notation for describing signs. The rule compiler checks the spreadsheets for missing entries, and if necessary adds new "blank" rows, using a model similar to that described in Sect. 2.

```
TrPhrase $$top
Source ce train ne circule pas via $$station
Target/gloss      TRAIN       CE          $$station      ALLER   PAS
Target/head       Down        Down        Neutral        Neutral Shaking
Target/gaze       Neutral     Down        Neutral        Neutral Neutral
Target/eyebrows   FurrowBoth  FurrowBoth  Up             Up      Neutral
Target/aperture   Small       Small       Neutral        Wide    Neutral
Target/mouthing   Tr@         SS          $$station      Vais    Pas
EndTrPhrase
```

```
TrLex $$station source="genève" gloss="GENEVE" mouthing="Genève"
```

Fig. 2. Examples of top-level translation rule and lexical entry for the train announcement domain. The rule defines a phrase of the form *ce train ne circule pas via ⟨station⟩* ("this train does not travel via ⟨station⟩". The lexical entry defines the translation for the name *genève* ("Geneva"). Only gloss and mouthing forms are defined for the lexical item.

3.3 Voice Questionnaires

We have already touched on the special problems of interactive voice questionnaires in the introduction. The overall intention is to add speech input and output capablities to the RAMP data gathering questionnaire framework [26]. The questionnaire definition encodes a branching sequence of questions, where the identity of the following question is determined by the answer to the preceding one. The display shows the person administering the questionnaire the field currently being filled; they formulate a question and speak it in their own language. In general, there are many questions which can be used to fill a given field, and the interviewer will choose an appropriate one depending on the situation. A basic choice, which affects most fields, is between a WH and a Y/N question. For example, if the interviewer can see recently used cooking utensils in front of him, it is odd to ask the open-ended WH-question "Where do you do the cooking?"; a more natural choice is to point and ask the Y/N confirmation question "Is cooking done in the house?"

As usual, the app performs speech recognition, gives the interviewer confirmation feedback, and speaks the target language translation if they approve. It then displays a question-dependent set of answer icons on the touch-screen. The respondent answers by pressing one of them; each icon has an associated voice recording, in the respondent language, identifying its function. Speech recognition coverage, in general, is limited to the specific words and phrases defined in the application content files. In this particular case, it is advantageous to limit it further by exploiting the tight constraints inherent in the questionnaire task, so that at any given moment only the subset of the coverage relevant to the current question is made available.

As far as rule formalisms are concerned, the questionnaire task only requires a small extension of the basic translation framework, in order to add the extra information associated with the questionnaire structure. The details are straightforward and are described in [1].

The next section uses an initial prototype of a voice questionnaire app ("AidSLT") to perform a simple evaluation of speech recognition performance. The questionnaire used for the evaluation contained 18 fields, which together supported 75 possible translated questions, i.e. an average of about 4 translated questions per field. The recognition grammar permitted a total of 11 604 possible source language questions, i.e. an average of about 155 source language questions per translated question.

4   Initial Evaluation

The initial AidSLT questionnaire was tested during the period March–July 2015 by seven humanitarian workers with field experience and knowledge of household surveys. The main focus of the evaluation was on the recognition of speech input by English-speaking interviewers. Subjects were presented with a simulation exercise that consisted in administering a household survey about malaria preventive measures to an imaginary French-speaking respondent. Instructions were sent by e-mail in the form of a PDF file. The subjects logged in to the application over the web from a variety of locations using password-protected accounts. Each subject ran the questionnaire once; annotations were added in the script so that several questions produced a popup which asked the subject to rephrase their initial question.

We obtained a total of 137 correctly logged interactions[7], which were annotated independently by two judges. Annotators were asked to transcribe the recorded utterances and answer two questions for each utterance: (a) whether the subject appeared to be reading the heading for the questionnaire field or expressing themselves freely, and (b) whether the translation produced adequately expressed the question asked in the context of the questionnaire task. Agreement between the two judges was very good, with a Cohen's kappa of 0.786 and an Intraclass Correlation Coefficient of 0.922.

The bottom-line result was that between 77 and 79 of the sentences were freely expressed (56–58 %) and only 10 produced incorrect translations (7 %), despite a Word Error Rate of about 29 %. All the incorrect translations were of course due to incorrect recognition. We find this result encouraging; the architecture appears to be robust to bad recognition and exploits the constrained nature of the task well.

5   Conclusions and Further Directions

We have described a platform that supports rapid development of a variety of limited-domain speech translation applications. Applications can be deployed on

---

[7] One subject misunderstood the instructions, one had severe audio problems with their connection, and a few utterances were spoiled by incorrect use of the push-to-talk interface.

the web and run on both desktop and mobile devices. The minimal formalism is designed to be used by domain experts who in general will not be computer scientists.

Although the translation platform is still at an early stage of development, experiences so far are positive; comparisons with our spoken CALL platform [25], which uses the same recognition architecture and has already been tested successfully on a large scale, leave us optimistic that we will achieve similar results here.

References

1. Armando, A., Bouillon, P., Rayner, E., Tsourakis, N.: A tool for building multilingual voice questionnaires. In: Proceedings of Translating and the Computer 36, London, England (2014)
2. Bouillon, P., et al.: Many-to-many multilingual medical speech translation on a PDA. In: Proceedings of the Eighth Conference of the Association for Machine Translation in the Americas, Waikiki, Hawaii (2008)
3. Braffort, A., Bossard, B., Segouat, J., Bolot, L., Lejeune, F.: Modélisation des relations spatiales en langue des signes française. Atelier LS, TALN, Dourdan, pp. 6–10 (2005)
4. Chang, D., Thyer, I., Hayne, D., Katz, D.: Using mobile technology to overcome language barriers in medicine. Ann. R. Coll. Surg. Engl. **96**(6), e23–e25 (2014)
5. Cox, S., Lincoln, M., Tryggvason, J., Nakisa, M., Wells, M., Tutt, M., Abbott, S.: Tessa, a system to aid communication with deaf people. In: Proceedings of the Fifth International ACM Conference on Assistive Technologies, pp. 205–212. ACM (2002)
6. Crasborn, O.: Nonmanual structures in sign language. In: Brown, K. (ed.) Encyclopedia of Language and Linguistics, vol. 8, 2nd edn, pp. 668–672. Elsevier, Oxford (2006)
7. Ebling, S., Glauert, J.: Exploiting the full potential of JASigning to build an avatar signing train announcements. In: Proceedings of the Third International Symposium on Sign Language Translation and Avatar Technology (SLTAT), Chicago, USA, October 2013, vol. 18, p. 19 (2013)
8. Ebling, S., Huenerfauth, M.: Bridging the gap between sign language machine translation and sign language animation using sequence classification. In: Proceedings of the 6th Workshop on Speech and Language Processing for Assistive Technologies (SLPAT) (2015)
9. Elliott, R., Glauert, J.R., Kennaway, J., Marshall, I.: The development of language processing support for the ViSiCAST project. In: Proceedings of the Fourth International ACM Conference on Assistive Technologies, pp. 101–108. ACM (2000)
10. Elliott, R., Glauert, J.R., Kennaway, J., Marshall, I., Safar, E.: Linguistic modelling and language-processing technologies for avatar-based sign language presentation. Univ. Access Inf. Soc. **6**(4), 375–391 (2008)
11. Forster, J., Schmidt, C., Koller, O., Bellgardt, M., Ney, H.: Extensions of the sign language recognition and translation corpus RWTH-PHOENIX-Weather (2014)
12. Fuchs, M., Tsourakis, N., Rayner, M.: A scalable architecture for web deployment of spoken dialogue systems. In: Proceedings of LREC 2012, Istanbul, Turkey (2012)
13. Huenerfauth, M.: Generating American Sign Language classifier predicates for English-to-ASL machine translation. Ph.D. thesis, University of Pennsylvania (2006)

14. Johnston, T., Schembri, A.: Australian Sign Language (Auslan): An Introduction to Sign Language Linguistics. Cambridge University Press, Cambridge (2007)
15. Kacorri, H., Lu, P., Huenerfauth, M.: Effect of displaying human videos during an evaluation study of American Sign Language animation. ACM Trans. Accessible Comput. **5**, 1–31 (2013)
16. Kennaway, R.: Synthetic animation of deaf signing gestures. In: Wachsmuth, I., Sowa, T. (eds.) GW 2001. LNCS (LNAI), vol. 2298, p. 146. Springer, Heidelberg (2002)
17. Marshall, I., Safar, E.: Grammar development for sign language avatar-based synthesis. In: Proceedings HCII, pp. 1–10 (2005)
18. Mazzei, A., Lesmo, L., Battaglino, C., Vendrame, M., Bucciarelli, M.: Deep natural language processing for Italian Sign Language translation. In: Baldoni, M., Baroglio, C., Boella, G., Micalizio, R. (eds.) AI*IA 2013. LNCS, vol. 8249, pp. 193–204. Springer, Heidelberg (2013)
19. Morrissey, S.: Data-driven machine translation for sign languages. Ph.D. thesis, Dublin City University, Dublin, Ireland (2008)
20. Morrissey, S., Way, A., Stein, D., Bungeroth, J., Ney, H.: Combining data-driven MT systems for improved sign language translation (2007)
21. Neidle, C.J., Kegl, J., MacLaughlin, D., Bahan, B., Lee, R.G.: The Syntax of American Sign Language: Functional Categories and Hierarchical Structure. MIT Press, Cambridge (2000)
22. Ong, S.C., Ranganath, S.: Automatic sign language analysis: a survey and the future beyond lexical meaning. IEEE Trans. Pattern Anal. Mach. Intell. **27**(6), 873–891 (2005)
23. Patil, S., Davies, P.: Use of Google Translate in medical communication: evaluation of accuracy. BMJ **349**, g7392 (2014)
24. Prillwitz, S., Leven, R., Zienert, H., Hanke, T., Henning, J.: HamNoSys: Version 2.0; Hamburg Notation System for Sign Languages; An Introductory Guide. Signum-Verlag (1989)
25. Rayner, M., Baur, C., Chua, C., Bouillon, P., Tsourakis, N.: Helping non-expert users develop online spoken CALL courses. In: Proceedings of the Sixth SLaTE Workshop, Leipzig, Germany (2015)
26. Salihu, M.: Piloting the first-ever RAMP survey using DHIS mobile. In: 141st APHA Annual Meeting, 2–6 November 2013. APHA (2013)
27. Stein, D., Schmidt, C., Ney, H.: Analysis, preparation, and optimization of statistical sign language machine translation. Mach. Transl. **26**, 325–357 (2012). Accessed 18 Mar. 2012
28. Su, H.Y., Wu, C.H.: Improving structural statistical machine translation for sign language with small corpus using thematic role templates as translation memory. IEEE Trans. Audio Speech Lang. Process. **17**(7), 1305–1315 (2009)
29. Tomaszewski, P., Farris, M.: Not by the hands alone: functions of non-manual features in Polish Sign Language. In: Bokus, B. (ed.) Studies in the Psychology of Language and Communication, pp. 289–320 (2010)
30. Tsourakis, N., Estrella, P.: Evaluating the quality of mobile medical speech translators based on ISO/IEC 9126 series: definition, weighted quality model and metrics. Int. J. Reliable Qual. E Healthc. (IJRQEH) **2**(2), 1–20 (2013)
31. Zeshan, U.: Sprachvergleich: Vielfalt und Einheit von Gebärdensprachen. In: Eichmann, H., Hansen, M., Heßmann, J. (eds.) Handbuch Deutsche Gebärdensprache, pp. 311–340. Signum, Hamburg (2012)

A Speech-to-Speech, Machine Translation Mediated Map Task: An Exploratory Study

Loredana Cerrato[1](✉), Hayakawa Akira[1], Nick Campbell[1], and Saturnino Luz[2]

[1] ADAPT Centre, School of Computer Science and Statistics,
Trinity College Dublin, Dublin, Ireland
{cerratol,campbeak,nick}@tcd.ie
[2] Usher Institute of Population Health Sciences and Informatics,
University of Edinburgh, Edinburgh, UK
S.Luz@ed.ac.uk

Abstract. The aim of this study is to investigate how the language technologies of Automatic Speech Recognition (ASR), Machine Translation (MT), and Text To Speech (TTS) synthesis affect users during an interlingual interaction. In this paper, we describe the prototype system used for the data collection, we give details of the collected data and report the results of a usability test run to assess how the users of the interlingual system evaluate the interactions in a collaborative map task. We use widely adopted usability evaluation measures: ease of use, effectiveness and users satisfaction, and look at both qualitative and quantitative measures. Results indicate that both users taking part in the dialogues (instructions giver and follower) found the system similarly satisfactory in terms of ease of learning, ease of use, and pleasantness, even if they were less satisfied with its effectiveness in supporting the task. Users employed different strategies in order to adapt to the shortcomings of the technology, such as hyper-articulation, and rewording of utterances in relation to error of the ASR. We also report the results of a comparison of the map task in two different settings – one that includes a constant video stream ("video-on") and one that does not ("no-video.") Surprisingly, users rated the no-video setting consistently better.

Keywords: Interlingual speech-to-speech translation · Repair strategies · Speaker alignment · Adaptation · User evaluation

1 Introduction

Speech-to-Speech (S2S) translation systems are becoming a daily reality as a way of communicating. Recently Microsoft announced and publicly demonstrated the Skype Translator: a system that enables cross-lingual conversations in real time. Although automatic S2S translation systems are in commercial deployment they are still not adapted to the way in which people actually behave when using them. To fill this gap we conducted an exploratory study using a prototype system to collect a set of 15 interactions in a S2S multimodal interlingual setting.

© Springer International Publishing Switzerland 2016
J.F. Quesada et al. (Eds.): FETLT 2015, LNAI 9577, pp. 53–64, 2016.
DOI: 10.1007/978-3-319-33500-1_5

We ran an evaluation study, during the collection of 15 interlingual dialogue interactions, in order to understand how the different language components together influence the interaction. A previous evaluation of a S2S Translation System, the NESPOLE! Project [9], has applied a multi-perspective approach, showing that the performance and usability of real-world S2S translation systems, are affected by several aspects and are not limited to the quality of the translation provided by the translation components within the system. Assessing the performance of the system's modules (ASR, MT and TTS) *per se*, is not the aim of this study. Here we focus on how users of the interlingual S2S system experience the overall interaction and try to understand how the different language technology components together influence the interaction. We also observe how user adaptation differs when users are provided with a live video stream of the remote participant.

2   Data Collection

By using a rapidly created S2S translation prototype system (ILMT-s2s system in Fig. 1) we recorded 15 interactions between speakers of different languages (English and Portuguese) who interacted remotely - over the network - to solve a specific task. The sessions were recorded in varying settings, but in this study we have divided them into the following settings:

No-video: Users communicate via voice to the ILMT-s2s system and only text of the interaction is displayed on the computer's screen – hence the interlocutors *cannot* see each other.

Video-on: Same setup as the "no-video" with an addition of a constant live video stream of the other user displayed on each interlocutor's computer screen – hence the interlocutors *can* see each other but can only hear the ILMT-s2s system output.

The ILMT-s2s system is able to record synchronised interaction data streams, such as: high quality audio, time-stamped ASR, MT and TTS events as well as biosignals[1] (heart rate, skin conductance, blood volume pressure and EEG). Also video of the interlocutor's actions were recorded from a camcorder placed above the computer screens and eye tracking glasses[2] or a wearable camcorder on either of the interlocutors. To provide a common topic of dialogue within all subjects, the dialogues were elicited using the HCRC Edinburgh Map Task technique [2]. While there are studies that used replications of the map task to look into communication in computer mediated tasks [12], this Interlingual Map Task is, to the best of our knowledge, the first corpus useful for the investigation of communicative behaviour in the presence of three additional "filters": ASR, MT, and TTS synthesis.

---

[1] Biosignal data is not used in this study.

[2] Eye tracking data is not used in this study.

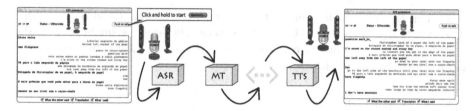

Fig. 1. Diagramatic representation of the S2S translation system (ILMT-s2s) used to collect the data (Color figure online)

The collected corpus is a rich source of data for analysis of different strategies for repairs and cooperative behaviour. It will be used to analyse several aspects of human adaptation to the technology and repair strategies. The data can also be directly compared with the original HCRC Map Task corpus to contrast interlingual and monolingual communication.

2.1 ILMT-s2s System

The ILMT-s2s system (refer to Fig. 1) was designed to minimise the user's action and to provide the user with visual feedback so as to facilitate the communication process. To talk to the interlocutor the user is to click and hold the "Push to talk" button while speaking and release it once the utterance is completed – similar to the user action for a walkie-talkie push-to-talk.

Once the "Push to talk" button is clicked, the text "Push to talk" changes to describe the current process being performed by the system. *"Recording..."*, *"Con. to flac"*, *"Con. to text"*, *"Translating."*, *"Sending text"*, and then back to *"Push to talk"*. Apart from the click and hold, no other user actions are required. For each utterance (by either participant in the dialogue), the ILMT-s2s system executes the following processes:

1. Record and save an audio file sampled at 96 kHz, 24 bit PCM format.
2. Down-sample the audio file to 16 kHz, 8 bit FLAC format and passes it on to the Google ASR interface.
3. Pass the result of the ASR to the MT service (Microsoft Bing) for translation.
4. Send the translated MT result to the remote participant's client computer.
5. TTS component converts the text to speech on the remote participant's client computer.

Feedback on the activities of the interlocutor is provided by the microphone icon displayed in the centre top of the display window, which turns orange when the participant clicks the "Push to talk" button. Similarly the colour of the small loud-speakers icons displayed beside the microphone turns orange when the TTS is outputting audio, as illustrated in Fig. 1.

A substantial part of the display window shows the ASR, MT and TTS results in text. The layout follows common text messaging applications. For the data collection, the MT results were not displayed[3], so the user could only see the ASR results and the text used for the TTS. When the ASR was unable to provide a result, an error message *"Error code (101); Please try again."* was displayed in-line, with no further details.

2.2   XML Record

The processes that the system executes are recorded in an XML file for each interlocutor. The file contains accurate time stamps, and descriptions of user and system events, including: utterance, ASR, MT and TTS process activation times and end times, and the outputs (parsed and raw) of these processes. This XML file also records the participant's details, source and target languages, etc.

2.3   Video Recordings

A frontal view of the subject is recorded in high-definition video from a Sony HDR-XR500 camcorder. These camcorders record the whole session and the video is used to analyse the state of the subject during the experiment and will be useful to annotate non-verbal information.

2.4   Audio Recordings

Three separate audio sources were recorded:

1. Hi-Resolution Audio (96 kHz, 24 bit) recording of the utterances passed on to the ASR process.
2. A standard audio recording from the HDR-XR500 camcorder's internal microphone. This is used to record the utterances that were not spoken to the system, i.e. muttering, sighs, and other reactions uttered by the participant.
3. A low quality audio recording from the eye tracking glasses and the bio-sensor software for data alignment.

The audio from the HDR-XR500 recording is used as a reference point to synchronise the other audio files. With the other audio files aligned to the HDR-XR500 recording, the time difference for the XML, eye tracking, bio-sensor data time log is determined. By aligning all the files related to one participant's experiment, the annotation data can be used with the various data sources (e.g., the annotation of the audio file can be aligned with the bio-sensor output as separately studied [5]).

2.5   Participants

Participants were recruited via an announcement on the Trinity College Dublin digital noticeboard and also via personal network connections. A total of 15

---

[3] It is interesting to note that, as indicated later in this paper under Sect. 3.1, the participants use the word "translation" to describe the ASR results or TTS output.

interactions of 15 native English speakers (♀5, ♂10), and 15 native Portuguese speakers (♀11, ♂4), between the ages of 18 to 45 were collected. The participants were instructed about the experiment via an information sheet which was sent to them at least 24 h before the expected time of data collection. At the time of the data collection they were given short verbal instruction on how to use the system. The recorded dialogues do not have a script, they are elicited according to the map task scheme, in which the two dialogue participants have a specific role designated *a priori*: an instruction giver and an instruction follower. The instruction giver has a map with a route drawn on it and has to instruct the follower to draw the same route on his/her unmarked copy of the map. The participants cannot see each other's map.

At the end of the recordings the participants were asked to complete a survey intended to assess the users' experience with the system.

The corpus is balanced as follows:

- 8 dialogues where the giver wears the bio-signal monitoring devices.
- 7 dialogues where the follower wears the bio-signal monitoring devices.
- 8 dialogues where interlocutors could see each other via video.
- 7 dialogues where interlocutors could not see each other via video.

Each recording session lasts between 20 and 74 min and contains between 33 and 199 utterances to the system. It was observed that the interlocutors with the role of giver produced an overall higher number of utterances corresponding to 30 %[4] more than interlocutors with the role of follower.

2.6 Synchronisation of the Recorded Data

Synchronisation of video and audio files were automatically performed using Final Cut Pro X from Apple Inc. Videos from both subjects were synchronised and cut accordingly into one 1080p video project and output as video in H.264 with 48 kHz 24 bit audio.

3 Data Transcription and Annotation

The recorded data was orthographically transcribed with the addition of some labels for interruptions, filled and empty pauses and noises. Transcription of the dialogues was carried out manually by two students (one native speaker of English and one native speaker of Portuguese) who listened to the audio-channel using Wavesurfer [14]. The two transcriber also carried out the annotation of the dialogues using the dedicated annotation tool ELAN [16]. They were trained by an expert annotator and were provided with specific guidelines and a sample of the annotation.

The annotation with ELAN was performed on a freely definable multi-layered (tracks) annotation scheme including the following tiers:

---

[4] Reduced to 17 % when outlier dialogue pair (giver : follower = 199 : 60) was removed.

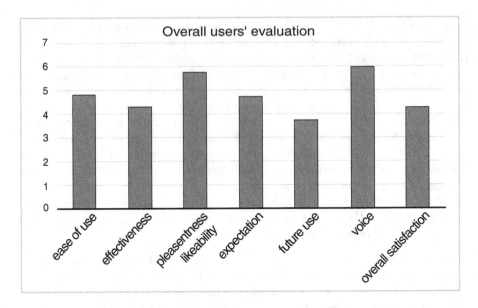

Fig. 2. Subjective user evaluation of the 15 interactions.

- Dialogue acts, following the annotation of [3] with a slight modification for the annotation of acknowledgments.
- Cognitive state: Surprised, Amused, Frustrated. These categories were motivated by the annotation schemes proposed by [7,13,15], but restricted to those three we thought were appropriate in this context.
- Facial expressions and head movements: Smile, Laughter, Surprise, Nods and Shakes. Inspired by the Mumin coding scheme [1], we chose the categories which we considered appropriate for our purposes.

Transcriptions and ELAN annotations alike were checked by two expert annotators to make sure that they complied with the guidelines. We therefore are confident that the data is reliable. For the annotation of the cognitive states we calculated the inter-coder agreement on one of the dialogues and the results are well above 60 %[5].

3.1   User Evaluation: Qualitative Study

After the interactions with the ILMT-s2s system, the participants were asked to complete a survey which aims at gathering a subjective assessment of the user's overall perception of the system. We used a 7 grade Likert scale to measure the level of agreement or disagreement on a number of given statements about the participants' perception of the system (e.g., Ease of use, Effectiveness and Satisfaction) (Fig. 2).

---

[5] Calculated using the modified kappa feature of ELAN 4.9.0's "Inter-Annotator Reliability..." function.

In general the participants rated the system as easy to learn and simple to use (average grade 4.9). Users found it pleasant and enjoyable (average grade 5.8), and they liked the voice of the TTS (this got the highest average grade in the evaluation: 6.0). When it comes to the overall satisfaction the average grade is 4.3. To the question related to the possible future use of the system the grade was 3.8. This might be due to the fact that the interactions were generally quite difficult, given the errors of the ASR and the MT. In fact, looking at the answers to the open questions in the evaluation form, with comments from the users about (a) adapting their speaking style to the system, (b) moments of irritations and (c) what was experienced as most difficult, the answers were consistent across the users. Generally the participants were conscious of the fact that they adapted their speech style to try to obtain better performance from ASR. The most common comments to the request *Please indicate why you changed the style of communication* was that they tended to speak in a slower manner, using simpler words and shorter utterances to try to improve ASR performance as can be read in these original comments from some of the users: *to make possible for the system to correctly translate I had to use simpler words and phrases; I spoke in shorter sentences to avoid bad speech recognition; I expected to be able to speak at my standard pace, but I couldn't. I had to speak in a slower pace so the system could understand some words; I had to change the style to fit the system: talk in a slower pace, try to articulate the words really well.* We observed several phenomena of adaptation in the interactions. Users tended, for instance, to adapt their speaking rate related to the level of recognition error [5].

When asked *indicate all the things that irritated you*, most of the users expressed their concern about the output of the automatic translation. Often mistranslations caused them to repeat some of the words over and over again (especially the names of the landmarks, which in fact are quite unusual and therefore caused difficulties to the MT system). Comments included: *Some words were not translated as I expected; Some translations bear no resemblance to what the speaker has input.* The mistranslations probably disrupted the flow of communication. However the unexpected and erroneous output of the translation was not always an element of irritation and frustration, since it often added an element of amusement and surprise to the task, as can be read in the following original comments: *I was not irritated at any points. I did laugh a few times though because I found that translating is something reasonably difficult and it was funny to see how pronunciation can impact so much on understanding; some of the misunderstanding of words were a bit annoying, but I thought that was funny.* Users were often surprised and amused by the mistranslations and observing their reactions will be very useful for our investigation of the affective states and cognitive load in this setting.

In all the sessions (except one) the participants managed to accomplish their task and carry out the communicative interaction, even if several errors occurred in the output of the ASR and MT.

Regarding the results when categorised by setting (video-on versus no-video) we notice that participants of no-video sessions (i.e. where they cannot see each

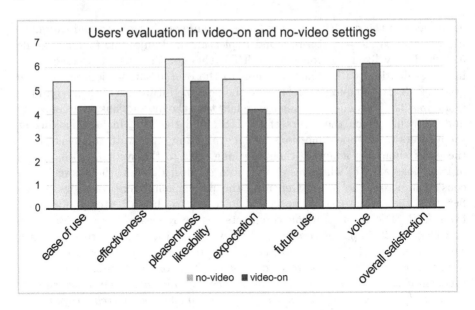

Fig. 3. Subjective user evaluation per givers and followers in the 2 settings.

other) seem to have had a better experience than participants of video-on sessions (i.e. when they can see each other) as illustrated in Fig. 3.

This result can be explained in different ways:

1. The network connection was not robust enough to support video transmission and this triggered system crashes (in 4 of the 8 interactions in the video-on setting there were between 1 and 4 system crashes). This might have slowed down the flow of interaction. In the 7 interactions in the no-video settings there were no systems crashes at all.
2. The possibility to see each other, without hearing each other, but with the ILMT-s2s system delay due to the strain on the computer caused by the video streaming process, might have added an element of distress or frustration. Two of the users in video-on sessions that suffered system crashes commented that the most irritating thing was the *lag time between input and translation*.
3. Dissatisfaction with the video-on setting might also stem from the fact that video raises expectations as regards to interactivity, that is: the participants are led to expect a fully interactive set but what they get is a mismatch between the real-time interaction on video and the chat-like, asynchronous interaction with the ILMT-s2s system.

In other words, having visual non-verbal feedback in real time and content exchange with a time delay disrupts communication more than the chat-only type of communication in the no-video setting, for which the participants had different expectations.

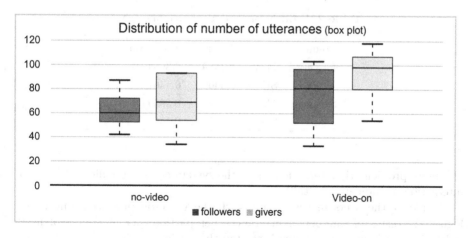

Fig. 4. Distribution of number of utterances per givers and followers in the two settings video-on and no-video.

Table 1. WER per setting

	no-video	video-on
Means	0.3781	0.4327
SD	0.6583	0.4499

3.2   User Evaluation: Quantitative Study

To check whether there was a different behaviour in the interactions in the two settings, video-on versus no-video, we looked at some quantitative measures for task effectiveness. Given the fact that the users had to click a button to talk, we counted the number of utterances distributed per giver and follower, which correspond to turns in our setting.

The results (Fig. 4) show that there is a increment of the number of utterances produced in the video-on setting. This can be interpreted in two different ways:

1. The higher number of utterances is an indication of a more fluent interaction and the video might have enhanced the interactive behaviour.
2. The higher number of utterances is an indication of a more problematic interaction due to a higher number of errors of the ASR or MT.

We therefore compared the Word Error Rates (WER) across the two settings. The WER is defined as the ratio of the Levenshtein distance between the aligned utterances (i.e. the number of additions, substitutions and deletions needed to convert one of the utterances into the other) to the number of words in the reference transcript [10] and the results show that the WER is indeed higher in the video-on setting as indicated in Table 1.

There also seems to be an increase in the WER after the system crashes, thus supporting the hypothesis that the higher number of utterances is an indication

Table 2. WER before and after system crash

video-on		w/ crash	w/o crash
Before crash	Means	0.4489	0.3954
	SD	0.4466	0.4467
After crash	Means	0.4821	0.3954
	SD	0.4553	0.4467

of a more problematic interaction, since the system crashes do affect the flow of interaction as indicated in Table 2.

Moreover the users in the setting with the video-on produced longer dialogues (in terms of number of utterances per participant). A couple of users (with the role of instruction giver), started the session by giving a very long set of instructions, which caused a failure of the ASR and hence a communication break-down. This also supports the hypothesis that the higher number of utterances is an indication of a more problematic interaction due to a higher number of errors of the ASR or MT.

3.3   User Evaluation: Discussion

A general assumption, supported by some previous studies, is that the video channel tend to enhance social aspects of communication [8], even if task effectiveness is mainly determined by the quality (in terms of low latency) of the audio channel, thus implying that video is of marginal importance [4]. In our interlingual setting we expected that the video channel would help to compensate for the misrecognition and translations errors, by providing a support channel for non-verbal communication.

This might still be the case in specific situations, and needs further analysis. Another design choice to study the use and usefulness of multi-modality in the context of multi-lingual communication would be to have a system that also records the video during the push-to-talk and then plays it at the same time as the translated speech. This way it might be possible to see if visual information (facial expressions, gestures) support the synthesised speech.

However in the set up of this study the results of the user evaluation show that communication in the video-on setting were seen as more problematic when compared to those in the no-video setting.

The results of the quantitative studies show that in the video-on sessions users tend to produce a higher number of utterances and the WER is higher compared to the no-video setting. This might be related to the pressure to be polite and friendly towards the other participant and create a more cooperative behaviour, which may not always be desirable in terms of effectiveness of task accomplishment. We observed that in the no-video sessions the participants tend to wait longer to receive information, expecting the other participant to speak. We speculate that this is because they cannot benefit from the visual information related to turn taking. This might suggest the need for a feedback indication like "over and out", that was used during the initial phases of radio wireless communication, to signal the actual end of a turn and yield the floor to the interlocutor.

It is possible that in S2S MT systems, which inevitably suffer from high latency due to processing constraints, the presence of the video channel has a disruptive rather than an enhancing effect on communication. This might be because the participants have a higher expectation of the system's performance. Previous findings on the benefits and limitations of a video channel on mediated interactions show that adding video to audio is more appropriate for group communication than one-to-one communication, and that adding video to audio is more effective for tasks involving strong personal relationships, which have an affective content. Video does not always improve the quality of the outcome, but it often improves the satisfaction of the partners [11]. In our setting, a one-to-one interaction, the users have to solve the map task and they did not know each other before the recordings, which means that they didn't have any personal relationship. This might also have been the reason why they experienced the video-on situation as problematic.

4   Future Work

The corpus collected in this study has been used to assess human reactions to an automated S2S translation system. Results from a study of how speakers adjust their speaking style in relation to errors from the ASR, while performing the Interlingual Map Task show that (a) system errors influence speaking rate, and (b) the perceived level of cooperation by the interlocutors increases as system error increases [6]. Another analysis of possible associations between speech recognition performance and three cognitive states that arise in dialogues mediated by a S2S machine translation system has been carried out [5].

Our plans for further research include detailed analysis of speech, video, eye movements, facial expressions and physiological signals on the data recorded. In addition to analysis of speech, gestures, and facial expressions, we intend to investigate possible correlations between biosignals and different communicative events (e.g. reaction to errors, surprise, etc.)

We expected that the knowledge acquired by analysing the data in this interlingual corpus can be used to provide baseline material for component development and testing and will also enable testing of methods for "affect sensing" from acoustic, video and biometric data recorded during the interaction.

Acknowledgments. This research is supported by Science Foundation Ireland through the CNGL Programme (Grant 12/CE/I2267) in the ADAPT Centre (www. adaptcentre.ie) at Trinity College, Dublin.

References

1. Allwood, J., Cerrato, L., Jokinen, K., Navarretta, C., Paggio, P.: The MUMIN coding scheme for the annotation of feedback, turn management and sequencing phenomena. Lang. Resour. Eval. **41**(3–4), 273–287 (2007)
2. Anderson, A.H., Bader, M., Bard, E.G., Boyle, E., Doherty, G., Garrod, S., Isard, S., Kowtko, J., McAllister, J., Miller, J., et al.: The HCRC map task corpus. Lang. Speech **34**(4), 351–366 (1991)
3. Carletta, J., Isard, S., Doherty-Sneddon, G., Isard, A., Kowtko, J.C., Anderson, A.H.: The reliability of a dialogue structure coding scheme. Comput. Linguist. **23**(1), 13–31 (1997)
4. Finn, K.E., Sellen, A.J., Wilbur, S.B. (eds.): Video-Mediated Communication. Lawrence Erlbaum Associates Inc., Hillsdale (1997)
5. Hayakawa, A., Cerrato, L., Campbell, N., Luz, S.: Detection of cognitive states and their correlation to speech recognition performance in speech-to-speech machine translation systems. In: Proceedings of INTERSPEECH 2015, pp. 2539–2543. ISCA, Dresden (2015)
6. Hayakawa, A., Cerrato, L., Campbell, N., Luz, S.: A study of prosodic alignment in interlingual map-task dialogues. In: Proceedings of ICPhS XVIII. No. 0760, University of Glasgow, Glasgow, United Kingdom (2015)
7. Henrichsen, P.J., Allwood, J.: Predicting the attitude flow in dialogue based on multi-modal speech cues. In: NEALT PROCEEDINGS SERIES (2012)
8. Kane, B., Luz, S.: Probing the use and value of video for multi-disciplinary medical teams in teleconference. In: Proceedings of CBMS 2006, pp. 518–523. IEEE Computer Society, Salt Lake City (2006)
9. Lavie, A., Metze, F., Cattoni, R., Costantini, E.: A multi-perspective evaluation of the NESPOLE!: speech-to-speech translation system. In: Proceedings of the ACL-02 Workshop on Speech-to-Speech Translation: Algorithms and Systems, vol. 7, pp. 121–128. Association for Computational Linguistics, Philadelphia (2002)
10. Levenshtein, V.I.: Binary codes capable of correcting deletions, insertions, and reversals. Sov. Phys. Dokl. **10**, 707–710 (1966)
11. Mariani, J.: Spoken language processing and multimodal communication: a view from Europe. In: Plenary Talk, NSF Workshop on Human-centered Systems: Information, Interactivity, and Intelligence (HCS), Arlington, VA, USA (1997)
12. Newlands, A., Anderson, A.H., Mullin, J.: Adapting communicative strategies to computer-mediated communication: an analysis of task performance and dialogue structure. Appl. Cogn. Psychol. **17**(3), 325–348 (2003)
13. Popescu-Belis, A.: Dialogue acts: one or more dimensions. ISSCO WorkingPaper 62 (2005)
14. Sjölander, K., Beskow, J.: Wavesurfer - an open source speech tool. In: Proceedings of INTERSPEECH 2000, pp. 464–467. ISCA, Beijing (2000)
15. The-AMI-Emotion-Annotation-Subgroup: Coding guidelines for affect annotation of the ami corpus. http://groups.inf.ed.ac.uk/ami/corpus/Guidelines/EmotionAnnotationManual-v1.0.pdf, no institute given
16. Wittenburg, P., Brugman, H., Russel, A., Klassmann, A., Sloetjes, H.: ELAN: a professional framework for multimodality research. In: Proceedings of LREC 2006, Genoa, Italy, pp. 1556–1559 (2006)

Combining Several User Models to Improve and Adapt the Dialog Management Process in Spoken Dialog Systems

David Griol[1(✉)], José Manuel Molina[1], Araceli Sanchis[1], and Zoraida Callejas[2]

[1] Computer Science Department, Carlos III University of Madrid, Leganes, Spain
{david.griol,josemanuel.molina,araceli.sanchis}@uc3m.es
[2] Department of Languages and Computer Systems,
University of Granada, Granada, Spain
zoraida@ugr.es

Abstract. Spoken dialog systems have demonstrated a high potential for more flexible, usable and natural human-computer interaction. These improvements are highly dependent on the users' adaptation and dialog management processes, which respectively integrates adaptation capabilities and decides the next system response for the current dialog state. In this paper we propose to carry out the fusion of the user's adaptation and dialog management processes at the dialog level in a single step. To do this, we describe an approach based on a statistical model that combines two models for user's intention modeling, generates a single representation from the users utterances and their confidence scores, and selects the next system action based on this representation. The paper also describes the practical application of the proposed approach to develop a dialog system providing travel and tourist information.

Keywords: Dialog management · User modeling · Spoken dialog systems · Statistical methodologies

1 Introduction

Speech and natural language technologies allow users to communicate in a flexible and efficient manner, making possible to access applications in which traditional input interfaces cannot be used (e.g. in-car applications, access for disabled persons, etc.). Also speech-based interfaces work seamlessly with small devices (e.g., smartphones and tablets PCs) and allow users to easily invoke local applications or access remote information. For this reason, spoken dialog systems [5,7,8] are becoming a strong alternative to traditional graphical interfaces which might not be appropriate for all users and/or applications.

However, spoken dialog systems developed for commercial or academic purposes are usually defined ad-hoc for their specific application domain and the adaptation capabilities of speech interfaces are frequently restricted to static choices [13]. For example, users have diverse ways of communication.

© Springer International Publishing Switzerland 2016
J.F. Quesada et al. (Eds.): FETLT 2015, LNAI 9577, pp. 65–76, 2016.
DOI: 10.1007/978-3-319-33500-1_6

Novice users and experienced users may want the interface to behave completely differently, such as maintaining more guided vs. more flexible dialogs. Processing context is not only useful to adapt the systems' behavior, but also to cope with the ambiguities derived from the use of natural language [11].

In this paper, we propose a framework to develop user-adapted spoken conversational agents. Our framework allows to dynamically incorporate user specific requirements and preferences to improve and personalize web information and services provided. The proposed framework is mainly focused on three specific processes carried out by dialog system: user adaptation, fusion of input information sources, and dialog management.

Research in techniques for user modeling has a long history within the fields of language processing and speech technologies. According to Zukerman and Litman [14], very early examples of user modeling in these fields are dominated by knowledge-based formalisms and various types of logic aimed at modeling the complex beliefs and intentions of agents [10]. In more recent years, dialog systems have tended to focus on cooperative, task-oriented rather than conversational forms of dialog, so that user models are now typically less complex. It is possible to classify the different approaches with regard to the level of abstraction at which they model dialog: acoustic level, word level, or intention-level.

Intention-level models are particularly useful to generate a compact representation of human-computer interaction. Intentions cannot be observed, but they can be described using the speech-act and dialog-act theories [12]. Two main approaches can be distinguished to the creation of user intention models: rule-based and data or corpus-based. In a rule-based user model, different rules determine the behavior of the system [6]. In this approach the researcher has complete control over the design of the evaluation study. However, these proposals are usually designed ad-hoc for their specific domain using models in which developers must specify each step to be followed by the user model.

Corpus-based approaches use probabilistic methods to generate the user input, with the advantage that this uncertainty can better reflect the unexpected behaviors of users interacting with the system. Statistical models of user intention have been suggested as the solution to the lack of the data that is required for training and evaluating dialog strategies.

As will be described in Sect. 2, our proposed user intention simulation technique is based on the combination of two models. The first model is based on a classification process that considers the complete dialog history by incorporating several knowledge sources, combining statistical and heuristic information to enhance the dialog model. The second model is focused on the simulation of the user and conversational agents to acquire a dialog corpus. At the beginning of the simulation, the set of system responses is defined as equiprobable. When a successful dialog is simulated, the probabilities of the answers selected by the conversational agent simulator during that dialog are incremented before beginning a new simulation.

Finally, the dialog management process has the main goal of selecting the next action of the system [3], interpreting the incoming semantic representation of the user input in the context of the dialog. Automating dialog management is useful for developing, deploying and re-deploying applications and also reducing the time-consuming process of hand-crafted design. The main trend in this area is an increased use of data for automatically improving the performance of the system and develop systems that exhibit more robust performance, improved portability, better scalability and easier adaptation to other tasks.

In this paper, we propose to merge the user modeling and dialog management processes by means of a statistical methodology that considers a set of input information sources provided by the spoken interaction with the user and the user's intention models, uses a data structure to store the values for the different input information sources received by the dialog manager along the dialog history, and selects the next system response by means of a classification process that takes this data structure as input.

The remainder of the paper is as follows. Section 2 presents our approach for developing user-adapted dialog systems. Section 3 describes the application of our approach to develop a practical system providing travel and tourist information. Section 4 presents the results of a preliminary evaluation of this practical dialog system. Finally, Sect. 5 presents the conclusions and suggests some future work guidelines.

2   Our Proposal to Develop User-Adapted Spoken Dialog Systems

Given the number of operations that must be carried out by a spoken dialog system, the scheme used for the development of these systems usually includes several generic modules that must cooperate to satisfy the user's requirements. The *Automatic Speech Recognition module* (ASR) transforms the user utterance into the most probable sequence of words. The *Spoken Language Understanding module* (SLU) provides a semantic representation of the meaning of the sequence of words generated by the ASR module. The *Dialog Manager* (DM) determines the next action to be taken by the system following a dialog strategy. The *Repository Query Manager* (RQM) receives requests for information or services, and returns the result to the dialog manager. The *Natural Language Generator module* (NLG) receives a formal representation of the system action and generates a user response in natural language. Finally, a *Text to Speech Synthesizer* (TTS) generates the audio signal transmitted to the user.

As explained in the introduction section, in our contribution, we want also to model the user intention as an additional valuable information source to be considered by the dialog manager. To do this, we propose the use of two models for modeling the user intention, which are explained in Subsects. 2.1 and 2.2. The outputs generated by the user models and the information provided by the user during the dialog are considered by the dialog manager to select the next system action as it will be explained in Subsect. 2.3.

2.1 First Method for Modeling the User Intention

The first methodology that we have developed for modeling the user intention extends our previous work in statistical models for dialog management [3]. Our proposed technique for user modeling simulates the user intention level by means of providing the next user dialog act in the same representation defined for the natural language understanding module. The lexical, syntactic and semantic information (e.g., words, part of speech tags, predicate-arguments structures, and named entities) associated to speaker u's ith clause is denoted as c_i^u.

Our model is based on the proposed in [1]. In this model, each user clause is modeled as a realization of a user action defined by a subtask to which the clause contributes, the dialog act of the clause, and the named entities of the clause. For speaker u, DA_i^u denotes the dialog label of the ith clause, and ST_i^u denotes the subtask label to which the ith clause contributes. The dialog act of the clause is determined from the information about the clause and the previous dialog context (i.e., k previous utterances) as shown in Eq. 1.

$$DA_i^u = \underset{d^u \in \mathcal{DA}}{\text{argmax}}\, P(d^u | c_i^u, ST_{i-1}^{i-k}, DA_{i-1}^{i-k}, c_{i-1}^{i-k}) \tag{1}$$

In a second stage, the subtask of the clause is determined from the lexical information about the clause, the dialog act assigned to the clause according to Eq. 1, and the dialog context, as shown in Eq. 2.

$$ST_i^u = \underset{s^u \in \mathcal{ST}}{\text{argmax}}\, P(s^u | DA_i^u, c_i^u, ST_{i-1}^{i-k}, DA_{i-1}^{i-k}, c_{i-1}^{i-k}) \tag{2}$$

In our proposal, we consider both static and dynamic features to estimate the conditional distributions shown in Eqs. 1 and 2. Dynamic features include the dialog act of each utterance and the task/subtask of each utterance. Static features include the words ans the part of speech tags in each utterance. As described in [1], the conditional distributions shown in Eqs. 1 and 2 can be estimated by means of the general technique of choosing the maximum entropy (MaxEnt) distribution that properly estimates the average of each feature in the training data. This can be written as a Gibbs distribution parameterized with weights λ as Eq. 3 shows, where V is the size of the label set, X denotes the distribution of dialog acts or subtasks (DA_i^u or ST_i^u) and Φ denotes the vector of the described static and dynamic features used for the user turns from $i-1 \cdots i-k$.

$$P(X = st_i | \phi) = \frac{e^{\lambda_{st_i} \cdot \phi}}{\sum_{st=1}^{V} e^{\lambda_{st_i} \cdot \phi}} \tag{3}$$

Each of the classes can be encoded as a bit vector such that, in the vector corresponding to each class, the ith bit is one and all other bits are zero. Then, V-one-versus-other binary classifiers are used as Eq. 4 shows.

$$P(y | \phi) = 1 - P(\overline{y} | \phi) = \frac{e^{\lambda_y \cdot \phi}}{e^{\lambda_y \cdot \phi} + e^{\lambda_{\overline{y}} \cdot \phi}} = \frac{1}{1 + e^{-\lambda_{\overline{y}}' \cdot \phi}} \tag{4}$$

where $\lambda_{\overline{y}}$ is the parameter vector for the anti-label $\overline{y}$ and $\lambda_{\overline{y}}' = \lambda_y - \lambda_{\overline{y}}$.

2.2   Second Method for Modeling the User Intention

The second method proposed for modeling the user intention is focused on the simulation of the user and conversational agents to acquire a dialog corpus. In our dialog generation technique, both agents use a random selection of one of the possible responses defined for the semantics of the task (expressed in terms of user and system dialog acts). At the beginning of the simulation, the set of system responses is defined as equiprobable. When a successful dialog is simulated, the probabilities of the answers selected by the conversational agent simulator during that dialog are incremented before beginning a new simulation.

One of the main problems which must be considered during the interaction with a conversational agent is the propagation of errors through the different modules in the system. The recognition module must deal with the effects of spontaneous speech and with noisy environments; consequently, the sentence provided by this module could incorporate some errors. The understanding module could also add its own errors (which are mainly due to the lack of coverage of the semantic domain). Finally, the semantic representation provided to the dialog manager might also contain certain errors. Therefore, it is desirable to provide the dialog manager with information about what parts of the user utterance have been clearly recognized and understood and what parts have not.

In our proposal, the user simulator provides the conversational agent with the semantic representation associated to the user input together with its confidence scores [2]. To do this, an error simulation agent has been implemented to include semantic errors in the generation of dialogs. This agent modifies the dialog acts provided by the user agent simulator once it has selected the information to be provided to the user. In addition, the error simulation module adds a confidence score to each concept and attribute in the semantic representation generated for each user turn.

For the study presented in this paper, we have improved this agent using a model for introducing errors based on the method described in [9]. The generation of confidence scores is carried out separately from the model employed for error generation. This model is represented as a communication channel by means of a generative probabilistic model $P(c, a_u | \tilde{a}_u)$, where a_u is the true incoming user dialog act, $\tilde{a}_u$ is the recognized hypothesis, and c is the confidence score associated with this hypothesis.

The probability $P(\tilde{a}_u | a_u)$ is obtained by Maximum-Likelihood using the initial labeled corpus acquired with real users and considers the recognized sequence of words w_u and the actual sequence uttered by the user $\tilde{w}_u$. This probability is decomposed into a component that generates a word-level utterance from a given user dialog act, a model that simulates ASR confusions (learned from the reference transcriptions and the ASR outputs), and a component that models the semantic decoding process.

$$P(\tilde{a}_u | a_u) = \sum_{\tilde{w}_u} P(a_u | \tilde{w}_u) \sum_{w_u} P(\tilde{w}_u | w_u) P(w_u | a_u)$$

Confidence score generation is carried out by approximating $P(c|\tilde{a}_u, a_u)$ assuming that there are two distributions for c. These two distributions are handcrafted, generating confidence scores for correct and incorrect hypotheses by sampling from the distributions found in the training data corresponding to our initial corpus.

$$P(c|a_w, \tilde{a}_u) = \begin{cases} P_{corr}(c) & if \quad \tilde{a}_u = a_u \\ P_{incorr}(c) & if \quad \tilde{a}_u \neq a_u \end{cases}$$

The conversational agent simulator considers that the dialog is unsuccessful when one of the following conditions takes place:

- the dialog exceeds a maximum number of system turns empirically determined for each specific application domain;
- the response selected by the DM corresponds to a query not made by the user simulator;
- the RQM module generates an error because the user model has not provided the mandatory data needed to carry out the query;
- the NLG module generates an error when the response selected by the DM involves the use of a data item not provided by the user model.

A user request for closing the dialog is selected once the conversational agent simulator has provided the information defined in its objective(s). The dialogs that fulfill this condition before the maximum number of turns are considered successful.

2.3   Fusion and Dialog Management Processes

In order to control the interactions with the user, our proposed statistical dialog management technique represents dialogs as a sequence of pairs (A_i, U_i), where A_i is the output of the dialog system (the system answer) at time i, and U_i is the semantic representation of the user turn (the result of the understanding process of the user input) at time i; both expressed in terms of dialog acts [3]. This way, each dialog is represented by:

$$(A_1, U_1), \cdots, (A_i, U_i), \cdots, (A_n, U_n)$$

where A_1 is the greeting turn of the system, and U_n is the last user turn. We refer to a pair (A_i, U_i) as S_i, the state of the dialog sequence at time i.

In this framework, we consider that, at time i, the objective of the dialog manager is to find the best system answer A_i. This selection is a local process for each time i and takes into account the previous history of the dialog, that is to say, the sequence of states of the dialog preceding time i:

$$\hat{A}_i = \underset{A_i \in \mathcal{A}}{\operatorname{argmax}} P(A_i|S_1, \cdots, S_{i-1}) \tag{5}$$

where set $\mathcal{A}$ contains all the possible system answers.

Following Eq. 5, the dialog manager selects the following system prompt by taking into account the sequence of previous pairs (A_i, U_i). The main problem to resolve this equation is regarding the number of possible sequences of states, which is usually very large. To solve the problem, we define a data structure in order to establish a partition in this space (i.e., in the history of the dialog preceding time i). This data structure, which we call *Interaction Register* (IR), contains the following information:

- sequence of user dialog acts provided by the user throughout the previous history of the dialog (i.e., the output of the NLU module);
- user dialog act predicted by the first model (generated by means of Eq. 1);
- user subtask predicted by the first model(generated by means of Eq. 2);
- user dialog act predicted by the second model (generated as described in Subsect. 2.2);

After applying these considerations and establishing the equivalence relation in the histories of dialogs, the selection of the best A_i is given by Eq. 6.

$$\hat{A}_i = \underset{A_i \in \mathcal{A}}{\operatorname{argmax}} P(A_i | IR_{i-1}, S_{i-1}) \tag{6}$$

We propose the use of a classification process to decide the next system action following the previous equation. From our previous work on dialog management [3], we propose the use of a multilayer perceptron for the classification, where the input layer receives the current state of the dialog, which is represented by the term (IR_{i-1}, A_i). The values of the output layer can be viewed as the a posteriori probability of selecting the different user intention given the current situation of the dialog.

3   Practical Application

We have applied our user-adaptation methodology to develop and evaluate an adaptive dialog system for a travel-planning domain. The system provides user-adapted information in natural language in Spanish about approaches to a city, flight schedules, weather forecast, car rental, hotel booking, tourist attractions, theater listings, and film showtimes. The information offered to the user is extracted from a web page that users can visually complete to incorporate additional information about a city already present in the system, update this information or add new cities. Different Postgress databases are used to store this information and automatically update the data that is included in the application. In addition, several functionalities are related to dynamic information (e.g., weather forecast, flight schedules) directly obtained from webpages and web services. Thus, our system provides speech access to facilitate travel-planning information that is adapted to each user taking context into account.

Semantic knowledge is modeled in our architecture using the classical frame representation of the meaning of the utterance. We defined eight concepts corresponding to the different queries that users can perform to the system (*City-Approaches, Flight-Schedules, Weather-Forecast, Car-Rental, Hotel-Booking,*

Tourist-Attractions, *Film-Show times*, and *Theater-Listings*). Three task-independent concepts have also been defined for the task (*Affirmation*, *Negation*, and *Not-Understood*). A total of 101 system actions (DAs) were defined taking into account the information that the system provides, requests or confirms.

Using the *City_Approaches* functionality, it is possible to know how to get to a specific city using different means of transport. If specific means are not provided by the user, then the system provides the complete information available for the required city. Users can optionally provide an origin city to try to obtain detailed information taking into account this origin. Context information taken into account to adapt this information includes user's current position, and preferred means of transport and city.

The *Flight_Schedules* functionality provides flight information considering the user's requirements. Users can provide the origin and destination cities, ticket class, departure and/or arrival dates, and departure and/or arrival hours. Using *Weather_Forecast* it is possible to obtain the forecast for the required city and dates (for a maximum of 5 days from the current date). For both functionalities, this information is dynamically extracted from external webpages. Context information taken into account includes user's current location, preferred dates and/or hours, and preferred ticket class.

The *Car_Rental* functionality provides this information taking into account users' requisites including the city, pick-up and drop-off date, car type, name of the company, driver's age, and office. The provided information is dynamically extracted from different webpages. The *Hotel_Booking* functionality provides hotels which fulfill the user's requirements (city, name, category, check-in and check-out dates, number of rooms, and number of people).

The *Tourist-Attractions* functionality provides information about places of interest for a specific city, which is directly extracted from the webpage designed for the application. This information is mainly based on users recommendations that have been incorporated in this webpage. The *Theatre_Listings* and *Film_Showtimes* respectively provide information about theater performances and film showtimes that takes into account the users requirements. These requirements can include the city, name of the theater or cinema, name of the show or film, category, date, and hour. This information is also considered to adapt both functionalities and then provide user-adapted information.

An example of the semantic interpretation of a user utterance using the list of described dialog acts described is shown in Fig. 1.

The *IR* defined for the task is a sequence of 57 fields, corresponding to:

- The eight possible queries that users can perform to the system (*City-Approaches*, *Flight-Schedules*, *Weather-Forecast*, *Car-Rental*, *Hotel-Booking*, *Tourist-Attractions*, *Theater-Listings*, and *Film-Showtimes*).
- A total of 45 possible attributes that users can provide to the system in order to generate a detailed response for the different queries (e.g., *Origin_City*, *Destination_City*, *Country*, *Departure_Date*, *Departure_Hour*, *Arrival_Date*, *Hotel_Name*, *Hotel_Category*, *Check_in_Date*, *Check_out_Date*, *Number_Rooms*, *Number_People*, *Category*, *Film*, *Cinema*, *Show*, *Theater*, etc.).

| **Input sentence:** |
| *Yes, I would like to know how to get to Valencia by car and which four stars hotels are available for tomorrow.* |
| **Semantic interpretation:** |
| (*Affirmation*) |
| (*City_Approaches*) |
| *City*: Valencia |
| *Means_Transport*: Car |
| (*Hotel_Booking*) |
| *City*: Valencia |
| *Hotel_Booking*: Car |
| *Category*: Four Stars |
| *Check_in_Date*: Tomorrow |

Fig. 1. An example of the labeling of a user turn in the travel-planning system

- Three task-independent concepts that users can provide (*Acceptance, Rejection* and *Not-Understood*).
- A reference to the predicted user response provided by the user intention recognizer.

A set of 150 scenarios were manually defined to cover the different queries to the system including different user requirements and profiles. Basic scenarios defined only one objective for the dialog; i.e. the user aims at obtaining information about only one type of the possible queries to the system (e.g., to obtain flight schedules from an origin city to a destination for a specific date). More complex scenarios included more than one objective for the dialog (e.g., to obtain information about how to get to a specific city, as well as car rental and hotel booking information).

4   Experiments

We have completed a preliminary evaluation of our proposal by means of a comparative assessment using a baseline system developing for the described task. Both systems integrate exactly the same modules, but the baseline system does not incorporates our proposed framework for user-adaptation in the dialog manager of the system. An initial corpus of 500 dialogs was acquired using the baseline system with real users [4]. This corpus has been used to develop the *User-adapted system*, which includes the described framework for modeling the user's intention and consider this information in the dialog management process.

A total of 300 additional dialogs were recorded from interactions of 20 users employing the Baseline and User-adapted systems. Each user acquired a total of 15 dialogs, 10 users interacted with the Baseline system and 10 users with the User-adapted system. The evaluation was carried out by students and lecturers in our department following the types of scenarios described in the paper in different settings with their own devices. An example of the defined scenarios is as follows:

```
User name: Javier Martín
Location: Mesones Street
Date and Time: 2012-06-05, 6:45pm
Device: SmartPhone 00-00-45-5A-02-D9
Objective: Science and literature cultural activities for today.
Listings for next weekend.
```

An objective and subjective evaluation were carried out. We considered the following measures for the objective evaluation:

1. Dialog success rate. This is the percentage of successfully completed tasks. In each scenario, the user has to obtain one or several items of information, and the dialog success depends on whether the system provides correct data (according to the aims of the scenario) or incorrect data to the user.
2. Average number of turns per dialog (nT).
3. Confirmation rate. It was computed as the ratio between the number of explicit confirmations turns (nCT) and the number of turns in the dialog (nCT/nT).
4. Average number of corrected errors per dialog (nCE). The average of errors detected and corrected by the dialog manager. We have considered only those which modify the values of the attributes and thus could cause the failure of the dialog. The errors are detected using the confidence scores provided by the ASR and NLU modules. Implicit and explicit confirmations are employed to confirm or require again values detected with low reliability.
5. Average number of uncorrected errors per dialog (nNCE). This is the average of errors not corrected by the dialog manager. Again, only errors that modify the values of the attributes are considered.
6. Error correction rate (%ECR). The percentage of corrected errors, computed as nCE/ (nCE + nNCE).

The results presented in Table 1 show that both systems could interact correctly with the users in most cases. However, the user-adapted system obtained a higher success rate, improving the baseline results by 9 % absolute. Using the baseline system, the average number of required turns is also reduced from 10.4 to 8.6 (significance value of 0.019). These values are slightly higher than the ones obtained by means of a simulated user model, as in some dialogs the real users provided additional information which was not mandatory for the corresponding scenario or asked for additional information not included in the definition of the scenario once its objectives were achieved.

The confirmation and error correction rates were also improved by the user-adapted system (significance value of 0.008), given that less information is required to the user, reducing the probability of introducing ASR errors. The main problem detected was related to user inputs misrecognized with a very high ASR confidence, and this erroneous information was forwarded to the dialog manager. However, as the success rate shows, this fact did not have a considerable impact on the system operation.

In addition, we asked the users to complete a questionnaire to assess their subjective opinion about the system performance. The questionnaire had five

Table 1. Results of the objective evaluation of the user-adapted and baseline systems with real users

	Success rate	nT	Confirmation rate	%ECR	nCE	nNCE
Baseline system	82 %	10.4	29 %	78 %	0.82	0.21
User-adapted system	91 %	8.6	26 %	87 %	0.89	0.14

questions: (i) Q1: *How well did the system understand you?*; (ii) Q2: *How well did you understand the system messages?*; (iii) Q3: *Was it easy for you to get the requested information?*; (iv) Q4: *Was the interaction rate adequate?*; (v) Q5: *Was it easy for you to correct the system errors?*. The possible answers for each one of the questions were the same: *Never, Seldom, Sometimes, Usually,* and *Always*. All the answers were assigned a numeric value between one and five (in the same order as they appear in the questionnaire). Table 2 shows the average results of the subjective evaluation.

Table 2. Results of the subjective evaluation of the baseline and user-adapted systems with real users (0=worst, 5=best evaluation)

	Q1	Q2	Q3	Q4	Q5
Baseline system	4.1	4.7	3.9	3.9	3.2
User-adapted system	4.4	4.8	4.6	4.5	3.5

From the results, it can be observed that both systems are considered to correctly understand the different user queries and obtain a similar evaluation regarding the facility of correcting errors introduced by the ASR module. However, the user-adapted system has a higher evaluation rate regarding the facility of obtaining the data required to fulfill the complete set of objectives of the scenario and the suitability of the interaction rate during the dialog.

5   Conclusions and Future Work

In this paper we have described a framework to develop user-adapted dialog systems. Using our framework it is possible to develop conversational interfaces that optimize interaction management and integrate different sources of information that make it possible for the application to adapt to the user and the context of the interaction. To show the pertinence of our proposal, we have implemented and evaluated a practical system that provides adapted tourist information to its users. The results show that the users were satisfied with the interaction with the system, which achieved high performance rates. We are currently using the framework to build applications in other increasingly complex domains implying different types of information and web services mashups.

Acknowledgements. This work was supported in part by Projects MINECO TEC 2012-37832-C02-01, CICYT TEC2011-28626-C02-02, CAM CONTEXTS (S2009/TIC-1485).

References

1. Bangalore, S., Fabbrizio, G.D., Stent, A.: Learning the structure of task-driven human-human dialogs. IEEE Trans. Audio Speech Lang. Process. **16**(7), 1249–1259 (2008)
2. García, F., Hurtado, L.F., Sanchis, E., Segarra, E.: The incorporation of confidence measures to language understanding. In: Matoušek, V., Mautner, P. (eds.) TSD 2003. LNCS (LNAI), vol. 2807, pp. 165–172. Springer, Heidelberg (2003)
3. Griol, D., Callejas, Z., López-Cózar, R., Riccardi, G.: A domain-independent statistical methodology for dialog management in spoken dialog systems. Comput. Speech Lang. **28**(3), 743–768 (2014)
4. Griol, D., García-Jiménez, M.: Development of interactive virtual voice portals to provide municipal information. Adv. Intell. Soft Comput. **151**, 161–172 (2012)
5. López-Cózar, R., Callejas, Z., Griol, D., Quesada, J.F.: Review of spoken dialogue systems. Loquens **1**(2), 1–15 (2014)
6. López-Cózar, R., de la Torre, A., Segura, J., Rubio, A.: Assessment of dialogue systems by means of a new simulation technique. Speech Commun. **40**, 387–407 (2003)
7. McTear, M., Callejas, Z.: Voice Application Development for Android. Packt Publishing, Birmingham (2013)
8. Pieraccini, R.: The Voice in the Machine: Building Computers That Understand Speech. MIT Press, Cambridge (2012)
9. Schatzmann, J., Thomson, B., Young, S.: Error simulation for training statistical dialogue systems. In: Proceedings of ASRU, Kyoto, Japan, pp. 526–531 (2007)
10. Schatzmann, J., Weilhammer, K., Stuttle, M., Young, S.: A survey of statistical user simulation techniques for reinforcement-learning of dialogue management strategies. Knowl. Eng. Rev. **21**(2), 97–126 (2006)
11. Seneff, S., Adler, M., Glass, J., Sherry, B., Hazen, T., Wang, C., Wu, T.: Exploiting context information in spoken dialogue interaction with mobile devices. In: Proceedings of IMUx 2007, pp. 1–11 (2007)
12. Traum, D.: Foundations of rational agency, chap. In: Speech Acts for Dialogue Agents, pp. 169–201. Kluwer (1999)
13. Whittaker, S.: Interaction design: what we know and what we need to know. Interactions **20**(4), 38–42 (2013)
14. Zukerman, I., Litman, D.: Natural language processing and user modeling: synergies and limitations. User Model. User-Adap. Inter. **11**, 129–158 (2001)

Exploring Random Indexing for Profile Learning

Adrian Fonseca Bruzón[1,2(✉)], Aurelio López-López[1], and José Medina Pagola[3]

[1] National Institute of Astrophysics, Optics and Electronics,
Sta María Tonantzintla, Puebla, Mexico
{adrian,allopez}@inaoep.mx
[2] Center for Pattern Recognition and Data Mining, Santiago de Cuba, Cuba
[3] Advanced Technologies Application Center, Havana, Cuba
jmedina@cenatav.co.cu

Abstract. Random Indexing is a recent technique for dimensionality reduction that allows to obtain a word space model from a set of contexts. This technique is less computationally expensive in comparison with others like LSI, Word2Vec or LDA. These characteristics turn it an attractive prospect to be used in an online learning environment. In this work, we compare several variants reported in the Random Indexing literature with the aim of using on the profile learning task. Experiments conducted in a subcollection of the dataset Reuter-21578 show that Random Indexing produces promising results, identifying some versions without actual advantage for the task at hand. Results obtained, by comparing Random Indexing with LDA, Word2Vec or LSI, also show that this technique is a viable alternative for representing documents.

Keywords: Random Indexing · Text Categorization · Profile learning

1 Introduction

Nowadays, personalization is a key component of many algorithms of Online Learning or Recommender Systems. Usually, these algorithms create a user profile for representing the user information needs. These algorithms have to decide for each document, whether it matches with the user profile or not.

In Text Mining, these methods are particularly important if we take into account the huge amount of new information that every day is generated on internet. However, in such tasks, these methods have to deal with two big problems, the language and the dimensionality. Natural language is a big challenge for computer science. On one side words are ambiguous, i.e. one word can have several meanings and several words can refer to the same concept. On the other hand, in the context of online learning, documents are continuously arriving, and usually they have some new unseen words that have to be taken into account later on.

The other problem is the dimensionality. Usually, documents are represented with a vector of a dimensionality equal to size of the vocabulary of the collection, or in a real environment equal to the total number of words found so far.

© Springer International Publishing Switzerland 2016
J.F. Quesada et al. (Eds.): FETLT 2015, LNAI 9577, pp. 77–85, 2016.
DOI: 10.1007/978-3-319-33500-1_7

High dimension causes more dispersion in document representation and imposes higher requirement for storing the documents. This situation significantly affects performance and quality of online learning algorithms.

Some algorithms have been reported in the literature with the aim of solving one or more of the aforementioned problems. Among them we can mention Latent Semantic Indexing (LSI) [4], Probabilistic Latent Semantic Indexing (PLSI) [7], Latent Dirichlet Allocation (LDA) [2] or Word2Vec[1] [10]. However, these methods are computationally expensive, or require access to the whole term-document frequency matrix during the semantic space construction. These disadvantages limit their incorporation into an online environment where frequent updates can occur in the information available.

Random Indexing can be an alternative, since this method is less expensive and does not require access to the whole term-document frequency matrix. For these reasons, this method is more attractive for use in an online environment. Moreover, several variants have been developed for Random Indexing intended for various tasks related to Natural Language Processing (NLP).

In this paper we report an experimental comparison of several of these variants on the context of profile learning. The results indicate that this representation can produce competitive results with a small dimension.

The rest of this paper is organized as follow: in the next section we describe Random Indexing and its major variants. Then we outline our proposal to use Random Indexing in profile learning. Next we present our experimental framework and discuss the results. Finally, we provide conclusions and possible areas for further work.

2   Random Indexing

Random Indexing [12,13,16] was introduced by Kanerva *et al.* in 2000 [9] and is based on three main assumptions:

- Distributional hypotesis: Words with similar meanings appear in similar contexts [15].
- Johnson – Lindenstrass Lemma: The projection of a high dimensional space into a space of much lower dimension can be done in such a way that distances between points are nearly preserved [8].
- And there exist many more pseudo-orthogonal directions than real orthogonals in a high dimensional space [5].

Kanerva's ideas were further developed by Magnus Sahlgren from the Swedish Institute of Computer Science. He formalized Random Indexing as a two step process [16] in the following way:

1. First, each context (e.g. a document or a word) in the data is assigned a unique and randomly generated representation called an index vector. These index vectors are sparse, high-dimensional, and ternary, which means that

---

[1] https://code.google.com/p/word2vec/.

their dimensionality (d) is on the order of thousands, and that they consist of a small number of randomly distributed +1s and −1s, with the rest of the elements of the vectors set to 0.

2. Then, context vectors are constructed by scanning through the text, and each time a word occurs in a context (e.g. in a document, or within a sliding context window), the context's d-dimensional index vector is added to the context vector for the word in question. Words are thus represented by d-dimensional context vectors that are effectively the sum of the words' contexts.

Different types of context can be used in the Random Indexing process. The most widely used are those considering the whole document as context or taking terms as context. When terms are considered as contexts, usually a window around the target term is taken into account. In that case, the context vector is updated with the index vectors of those terms that are found in the neighborhood of the target term.

Another approach for considering terms as context was presented in [11]. In this work, an index vector is assigned to every term. In this case, the context vector is updated with all the index vectors of those terms found in the document.

Random Indexing captures the term semantics based on co-occurrences. However, Cohen *et al.* conclude that this technique has some shortcomings for determining indirect relations between words [3]. To overcome such limitation, they propose an extension named Reflective Random Indexing. In this extension, they assign an index vector to every term, then a representation for a document d is obtained as the sum of the index vectors of those terms that appear in d. Thereafter, these document vectors are employed to build the term context vector. This process can be repeated several times, but in their experiments, the best results are obtained after one or two iterations.

After obtaining the context vectors, we can obtain the representation of a document d, aggregating the context vectors of those terms that appear in it [17]. During the aggregation process, context vectors can be multiplied by a weight indicating the importance of each term in the document.

During the process of obtaining the final representation for a document, we can perform some transformation over the context vectors; in particular Higging & Burstain propose to substract the mean context vector from every context vector before obtaining the document representation [6]. According to the authors, in Random Indexing the similarity between documents is bound to increase with their length, and regardless of their relatedness. With this transformation, they try to mitigate this drawback.

3 Profile Learning with Random Indexing

Most of the works reported in the literature have used the traditional Bag of Words (BOW) model for representing documents. However, it is known that this model can not capture those semantic relations which exist between terms in a document [1].

On the other hand, several tasks like Information Filtering, News Recommendation, Text Categorization and in particular Profile Learning can benefit from employing some method that does not assume that the terms of a document are independent.

To our knowledge, it has not been explored yet the possibilities of Random Indexing on profile learning. In this scenario, Random Indexing is a plausible representation, having the advantage of being less expensive than other techniques, such as LSI, PLSI, LDA or Word2Vec.

In this study, we explore diferent variants reported in the Random Indexing literature in the context of profile learning. Profiles are internal representations of some information needs or interests. Usually, this task is modelled as a binary classification process where the classifier has to decide for each document whether it matches or not with every different profile.

A simple profile representation is created aggregating in a single vector all those documents that are relevant to the user. Following the same idea, we can build a vector that represents the irrelevant information.

With this representation, every new document is classified as relevant for the user if the similarity with the relevant profile vector is higher than that obtained with respect to the irrelevant profile vector.

When we are employing a semantic model like Random Indexing, an extra step is required. During training, it is necessary to consider in the profile all the available information from the training collection. This semantic model will be used for representing both training documents and the new ones to be classified.

4   Experiments

For the experimentation, we used the Reuters-21578[2] test collection. Several subsets have been created from this dataset, the most well known are:

- the set of 10 categories with the highest number of positive training examples.
- the set of 90 categories with at least one positive training example and one test example.
- the set of 115 categories with at least one positive training example.

In particular, for this study we selected the first subset. Table 1 shows the number of training documents for each class.

To avoid the class imbalance problem, we discarded the class *earn* and *acq*, considering in this study the remaining eight classes. Note that classes *earn* and *acq* account for more than 60 % of the objects in the subset.

During document preprocessing, tags and stop words were removed and a lemmatization process was applied. A TF-IDF term weigthing schema was computed.

Experiments were conducted to compare the performance obtained with several variants of Random Indexing. To this end, we performed a 5-fold cross-validation.

---

[2] http://www.daviddlewis.com/resources/testcollections/reuters21578/.

Table 1. Statistics of Reuters-21578(10) Subset

Class	Number of training documents
Earn	3753
acq	2131
Wheat	264
Money-fx	600
Corn	206
Trade	449
Grain	527
Interest	389
Crude	510
Ship	276

For each class, the profile was represented as two vectors. One representing those documents relevant for the user and the other for the non relevant. Each was created adding all vectors that belong, or not, to the class in the training set.

During the classification process, a document was tagged as Relevant to a profile if its similarity to the vector representing the relevant documents is higher than that obtained with respect to the non-relevant documents.

As evaluation measure, the traditional *precision* was selected, i.e. the proportion of documents classified as relevant that indeed are relevant, and also the *recall* measure, i.e. the proportion of relevant documents that are indeed classified as relevant. These measures usually are combined in the popular F_1 measure, $F_1 = \frac{2*precision*recall}{precision+recall}$.

Since F_1 measure is computed separately for each class, we considered as global measure the mean over all classes, commonly known as $Macro - F_1$.

In Fig. 1, we show the average obtained by $Macro - F_1$. In Table 2, we also include the average obtained for $Macro - Precision$ and $Macro - Recall$.

In the figure, RI represents Random Indexing when documents are considered as contexts; in the same way, wRI when terms are considered as context and a window around the target term is used and TRI when no window is used. RRI refers to Reflective Random Indexing. Those models with the suffix "-MV" represent the results obtained when mean context vector is subtracted from context vector before reaching the document representation.

For Random Indexing, we consider vectors of size 5000, with 5 position set as 1 and 5 positions defined as -1, when generating the index vectors. For wRI models we consider a window of size two around the word. For RRI only 1 iteration was done.

Also, we present the results obtained when we apply more traditional techniques like LSI or LDA. In LSI we consider 200 topics as parameter, and for LDA 1000. For Word2Vec, it was selected a size of 500, with a window of 7 elements and no word was ignored. All models were trained considering only the

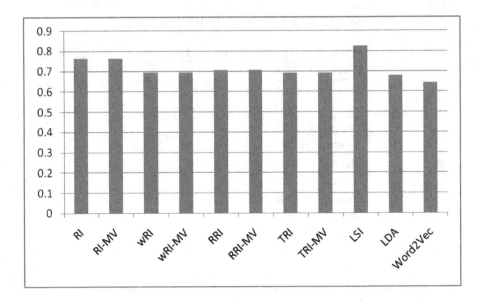

Fig. 1. Average of $Macro - F_1$

samples of the training data. In particular, for these techniques we employed the implementation provided by gensim[3] [14].

5 Discussion

From the results we can notice several behaviors. First, when the different techniques of Random Indexing are compared, the best results are obtained when documents are considered as contexts. In this case, this variant is better with respect to the other Random Indexing models in around 8% – 9%.

Other relevant aspect is that the results obtained for Reflective Random Indexing are better than the results obtained with Random Indexing when terms are considered as context. Reflective Random Indexing was intended to capture indirect relations between terms, and its usefulness is not the same in other tasks.

We can observe that, in most cases, when we subtracted the mean context vector from context vectors before obtaining the final document representation no consistent gains are reached. So, we do not find a reason to justify this additional operation without an actual improvement.

On the other hand, Random Indexing obtained similar results to techniques most widely used in text mining, such as LSI and LDA. Even, in our experiments, Random Indexing produced an outcome quite close to LSI (practically similar Macro-Recall but lower Macro-Precision) and better results than those obtained by LDA or Word2Vec.

---

[3] https://radimrehurek.com/gensim/.

Table 2. Average of $Macro - F_1$ for several variants of RI compared against LSI, LDA and Word2Vec

Variant	Averaged $Macro - Prec.$	Averaged $Macro - Recall$	Averaged $Macro - F_1$
RI	0.684	0.895	0.762
RI-MV	0.684	0.895	0.762
wRI	0.594	0.883	0.696
wRI-MV	0.594	0.883	0.696
RRI	0.613	0.865	0.706
RRI-MV	0.613	0.865	0.706
TRI	0.595	0.866	0.693
TRI-MV	0.594	0.863	0.692
LSI	0.783	0.9	0.825
LDA	0.663	0.768	0.680
Word2Vec	0.528	0.868	0.644

Finally, the main advantage of Random Indexing is that we only consider vectors of 5000 elements. This aspect takes a particular importance when we aim for online environments, where every new document can contain new terms. With Random Indexing, the problem of frequent new terms does not affect the efficiency since documents are always represented with vectors of fixed length.

When we are working in an online environment, where the training set grows up frequently, Random Indexing has an additional advantage over other methods, like LSI. Specifically, this advantage consists on the fact that with Random Indexing we do not requiere to keep previous documents when adding new information to the model, unlike LDA, LSI or Word2Vec. Also, Random Indexing is less expensive, since its processing involves only simple vector operations as addition, while the other techniques employ Singular Value Decomposition (SVD) or other optimization methods.

6 Conclusions

Random Indexing is an indexing technique that implicitly includes a dimensionality reduction, and in a simple iterative process can collect the semantic relations that exist among terms. Several approaches are reported in the literature for Random Indexing, applied to various different tasks. In this paper we report a comparison of the most relevant versions of Random Indexing intended for the profile learning task. The results reported show that considering documents as context achieves the best result, although with vector of approximately one third of the size of the full term-dimensionality. We also showed that Random Indexing produces slightly lower results than LSI but in less time and a third of the space. According to our experiments, Random Indexing can even achieve better results than those obtained by LDA or Word2Vec, with less space resources.

Despite these encouraging results, it remains to analyse the effect of the size of the experimental database; considering that in a semantic model, many documents are necessary to reach a valid representation of the actual relations that there exist among terms.

In that direction, in future works we will analyse the behavior of Random Indexing in relation with the amount of data available for its construction. Also, we plan to evaluate how the class imbalance problem affects the behavior of Random Indexing.

Acknowledgments. We thank the CPU Lab for the use of its facilities for running the experiments.

The first author was supported by Conacyt through scholarship 635046, and the second author was partially supported by SNI, México.

References

1. Becker, J., Kuropka, D.: Topic-based vector space model. In: Proceedings of the 6th International Conference on Business Information Systems, pp. 7–12 (2003)
2. Blei, D.M., Ng, A.Y., Jordan, M.I.: Latent dirichlet allocation. J. Mach. Learn. Res. **3**, 993–1022 (2003)
3. Cohen, T., Schvaneveldt, R., Widdows, D.: Reflective random indexing and indirect inference: a scalable method for discovery of implicit connections. J. Biomed. Inform. **43**(2), 240–256 (2010). http://www.sciencedirect.com/science/article/pii/S1532046409001208
4. Dumais, S., Furnas, G., Landauer, T., Deerwester, S., Deerwester, S., et al.: Latent semantic indexing. In: Proceedings of the Text Retrieval Conference (1995)
5. Hecht-Nielsen, R.: Context vectors: general purpose approximate meaning representations self-organized from raw data. In: Computational Intelligence: Imitating life, pp. 43–56 (1994)
6. Higgins, D., Burstein, J.: Sentence similarity measures for essay coherence. In: Proceedings of the 7th International Workshop on Computational Semantics, pp. 1–12 (2007)
7. Hofmann, T.: Probabilistic latent semantic indexing. In: Proceedings of the 22nd annual international ACM SIGIR Conference on Research and Development in Information Retrieval, pp. 50–57. ACM (1999)
8. Johnson, W.B., Lindenstrauss, J.: Extensions of lipschitz mappings into a hilbert space. Contemp. Math. **26**(189–206), 1 (1984)
9. Kanerva, P., Kristofersson, J., Holst, A.: Random indexing of text samples for latent semantic analysis. In: Proceedings of the 22nd Annual Conference of the Cognitive Science Society, vol. 1036 (2000)
10. Mikolov, T., Chen, K., Corrado, G., Dean, J.: Efficient estimation of word representations in vector space. In: Proceedings of Workshop at ICLR (2013)
11. Musto, C.: Enhanced vector space models for content-based recommender systems. In: Proceedings of the Fourth ACM Conference on Recommender Systems, RecSys 2010, pp. 361–364. ACM (2010)
12. QasemiZadeh, B., Handschuh, S.: Random indexing explained with high probability. In: Král, P., et al. (eds.) TSD 2015. LNCS, vol. 9302, pp. 414–423. Springer, Heidelberg (2015). doi:10.1007/978-3-319-24033-6_47

13. QasemiZadeh, B.: Random indexing revisited. In: Biemann, C., Handschuh, S., Freitas, A., Meziane, F., Métais, E. (eds.) NLDB 2015. LNCS, vol. 9103, pp. 437–442. Springer, Heidelberg (2015)

14. Řehůřek, R., Sojka, P.: Software framework for topic modelling with large corpora. In: Proceedings of the LREC 2010 Workshop on New Challenges for NLP Frameworks, ELRA, Valletta, Malta, pp. 45–50, May 2010 http://is.muni.cz/publication/884893/en

15. Rubenstein, H., Goodenough, J.B.: Contextual correlates of synonymy. Commun. ACM 8(10), 627–633 (1965)

16. Sahlgren, M.: An introduction to random indexing. In: Proceedings of the Methods and Applications of Semantic Indexing Workshop at the 7th International Conference on Terminology and Knowledge Engineering, TKE 2005, August 2005

17. Sahlgren, M., Cöster, R.: Using bag-of-concepts to improve the performance of support vector machines in text categorization. In: Proceedings of the 20th International Conference on Computational Linguistics, pp. 487. Association for Computational Linguistics (2004)

Text Categorisation
by Using Sentiment Composition

Diego Uribe[✉]

División de Estudios de Posgrado e Investigación, Instituto Tecnológico de la Laguna,
Revolución y Cuauhtémoc, Torreón, Coahuila, Mexico
diego@itlalaguna.edu.mx

Abstract. In recent years there has been an increasing interest in techniques for dealing with the compositionality of meaning. In fact, to derive the meaning of complex expressions (i.e. phrases and sentences) from the meanings of their parts has grabbed the researchers' attention. In this paper, we examine semantic composition from the perspective of sentiment composition: if the meaning of a sentence is a function of the meanings of its parts, the polarity of a sentence is a function of the polarities of its parts. Basically, we propose a model based on sentential sentiment composition in order to categorise a text review (i.e. an opinion) according to the polarity of the sentences it contains. The experimental results showed that our approach is a plausible alternative to categorise subjective texts.

1 Introduction

Capturing the meaning of linguistic expressions has always appealed to researchers. Syntax-driven semantic analysis, the process whereby formal representations are created to capture such meaning, is based on the Principle of Compositionality described by Partee et al. [1]:

The meaning of a complex expression is a function of the meaning of its parts and the syntactic rules by which they are combined.

In other words, the meaning of linguistic expressions is based on the knowledge of the lexicon and the grammar. Moreover, it is important to remark on how the principle of compositionality has been applied in the task of sentiment classification. In fact, Molainen and Pulman show how the classification of a complex constituent is a function of the classification of its sub constituents [2]. Specifically, they argue that, as far as sentential sentiment level is concerned, if the meaning of a sentence is a function of the meaning of its parts, then the polarity of a sentence is also a function of the polarity of its parts.

Grounded on these relevant concepts, we investigate in this work how plausible is to make use of sentiment composition to categorise a text review (i.e. an opinion) according to the polarity of the sentences it contains. In other words,

© Springer International Publishing Switzerland 2016
J.F. Quesada et al. (Eds.): FETLT 2015, LNAI 9577, pp. 86–95, 2016.
DOI: 10.1007/978-3-319-33500-1_8

we analyse the use of sentential sentiment composition for document-level classification. From this perspective, and taking into account the principle of compositionality, we want to explore at what extent the polarity of a particular review may be determined as some function of the polarities of its subconstituents, that is, its sentences.

Since a text review is composed of a set of sentences where each one expresses either objective or subjective information, each sentence is commonly classified as non-neutral, that is, sentences that denote a particular polarity, and neutral sentences, that is, impartial sentences. To illustrate the diversity of information expressed in the sentences that compose a particular text review, consider the following fragment extracted from a Movie review:

"I found it to be hysterical, as well as touching. My boyfriend doesn't even like Christmas and told me he had goosebumps for the last 20 min of the movie.
Is there a lesson to be learned? Does it matter? Christmas is about feel-good movies and when do you feel better then when you are laughing."

There are five sentences in the example, from which the first sentence has been classified as positive, the second one as negative, the third and fourth as neutral, and the last one as positive. Such classification has been carried out by using the Stanford CoreNLP toolkit [3], the sentiment classifier that we use in our experimentation, and which works predicting five sentiment classes: Very Positive, Positive, Neutral, Negative and Very Negative.

In this paper, the linguistic challenge that concerns our investigation is to categorise a text review represented as a set of sentences where each one expresses a particular polarity. In fact, we propose a sentiment composition model based on the analysis of the impact of each sentence and its contribution to the final emotion orientation of the text review. In other words, the classification of the review is determined as a function of the propagation of the sentiment of the sentences that it contains.

In our sentiment composition model, that we describe in Sect. 3, we calculate in a systematic way the polarity of the review as a function of the impact of its subconstituents: each sentence and its contribution to the final emotion orientation of the review. In other words, the classification of the review is determined as a function of the propagation of the sentences' sentiment that it contains. The experimentation conducted is described in Sect. 4 where the obtained results show how our approach is a plausible alternative to categorise subjective texts by making use of sentential sentiment composition. But we now go to the next Sect. 2 that makes a brief description of related work on sentiment composition.

2 Related Work

In this section, we briefly describe some of the substantial works dealing with the challenge of semantic composition. On the one hand, we make reference to works concerned with the integration of Distributional Semantics in the investigation of semantic composition. Distributional Semantics is predicted on the assumption that two words are semantically similar to the extent that their contextual representations are similar [4], so diverse computational models make use of this assumption as an experimental framework for semantic analysis. To determine the semantic similarity between linguistic expressions, the definition of distributional vectors play a crucial role as the contextual representations of such linguistic expressions.

Mitchell and Lapata consider semantic composition as a function of two distributional vectors that denote either phrases or sentences [5,6]. To determine the semantic similarity of phrases or sentences, they propose two functions as the operations to be conducted on the distributional vectors: additive and multiplicative functions. The experimental results showed how the composition models based on multiplicative functions capture semantic similarity more accurately.

Another interesting work was conducted by Guevara [7]. This work is basically an extension of the additive model of Mitchell and Lapata (mentioned lines above), where the main difference lies in the distributional vectors to be extracted from the corpus. Rather than extracting two distributional vectors in order to obtain the composed vector by using any of the functions proposed, Guevara's approach extracts three distributional vectors: the additional vector denotes the contextual representation of the target linguistic expression. In this way, the focus of Guevara relies on the use of a supervised learning machine technique (multivariate regression) whose purpose is to optimise the parameters (i.e. weights) of the composition model or function.

On the other hand, we make reference to the work concerned with the integration of semantic composition in the investigation of sentiment classification. As we previously mentioned, Moilanen and Pulman claim that if the meaning of a sentence is a function of the meaning of its parts, then the polarity of a sentence is also a function of the polarity of its parts [2]. The sentiment composition model requires two constituents (subconstituents) to estimate the polarity for the composite constituent (larger syntactic constituent). The composition operations to be implemented on the syntactic constituents are basically three: sentiment propagation (compositions where the polarity of a neutral constituent is override by the polarity of a non-neutral constituent), polarity reversal (compositions where the polarity of a non-neutral constituent is reversed) and polarity conflict resolution (disambiguating compositions where a polarity conflict occurs).

3   Composition Model

We describe in this section our semantic composition model. As we previously mentioned, the motivation of this research is to investigate the use of sentential sentiment composition for document-level classification. Thus, instead of viewing a document as a collection of terms, we contemplate a document as a collection of sentences. We next formally describe our task:

Given a set of sentiment classes C denoting the intensity of emotion expressed in a particular sentence, and a set of opinionated documents D, where each document $d \in D$ contains a set of possibly opinionated sentences S, and each sentence $s \in S$ is represented by a particular sentiment class $c \in C$, determine the sentiment class c of the whole document d.

Once the problem has been defined, and before describing our model, we can consider intuitive mechanisms to determine the sentiment class or polarity of a text review. By using measures of central tendency we can take a glance at the distribution of the polarities of the sentences that a review contains. For example, by using the mode, the most frequent sentiment class in the set of sentences may determine the polarity of the review. Likewise, by using the mean, the average sentiment corresponding to the set of sentences may also determine the polarity of the review. However, the problem of measures of central tendency is that we don't know to what extent the use of a single value represents the distribution of the multiple polarities corresponding to a set of sentences: the polarity values in the distribution can be either similar or different from the mean.

Our semantic composition model is based on the assessment of the strength of the sentiment expressed in each sentence. Said in another way, our model is represented as a function of the weight associated to each sentiment class. A formal representation is:

$$p(d) = f(w, c) \tag{1}$$

where $p(d)$ denotes the polarity of a particular text review d, and w denotes a specific weight associated to a particular sentiment class $c \in C$.

Now, to define the function entails to describe the composition operation to be implemented on the pairs that represent the weight associated to each sentiment class. Since the sentiment classification of each sentence is available, we can accumulate the impact or contribution of each sentence to the final orientation of the whole document. In other words, the sentiment propagation function is defined as follows:

$$f = \sum_{i=1}^{n} (w_i, c_i) \tag{2}$$

Basically, our composition model is a *weighted additive model*. It is important to notice that the similarity to the model proposed by Lapata [5] lies on the name only. In our case, instead of adding distributional vectors, the strengths of the sentences, represented by weighted values, is accumulated. In fact, by adding the sentences' polarities the propagation of the sentiment is being mainly guided by the impact of non-neutral sentences. For this reason, the influence of neutral sentences is meaningless.

4   Experimental Evaluation

The experimentation conducted is detailed in this section. First, we describe the type of documents to be analysed, and then, the linguistic processing implemented on the documents. The section concludes with the setting of the parameters of the composition model.

4.1   Corpus

Since the type of texts to be categorised is the document commonly known as opinion or review, the corps used in our experimentation consists of opinions grouped by a particular domain. The corpus is a collection of Epinions reviews developed by Taboada et al. [8] and consists of eight different domains: books, cars computers, cookware, hotels, movies, music and phones. There are 50 opinions per domain, giving a total of 400 reviews in the collection, which contains a grand total of 279,761 words. Also, each domain contains 25 reviews per polarity that denotes to have a balanced dataset per domain.

For our purpose, we only consider those domains with at least 4,000 terms. Thus, we discarded two domains: Cookware and Phones. These domains have been discarded because we can consider them as *outliers*: their number of terms is clearly separated from the rest of the domains. Then, from the remaining six, four domains were finally selected: Books, Computers, Movies and Music.

Additional information about the corpus is the specification of the total number of sentences corresponding to each domain. Since the basic unit of information to be analysed in our model is the sentence, rather than phrases or words, we show in Table 1 the total number of sentences corresponding to each domain as well as the average number of sentences per text review.

Table 1. Number of sentences per domain

Domain	Sentences	Sentences/Review
Books	1,589	32
Computers	2,669	53
Movies	1,796	36
Music	2,720	54

4.2   Pre-processing

Each review in the collection was submitted to linguistic analysis by using an integrated framework of linguistic tools: the Stanford CoreNLP toolkit [3]. In fact, Stanford CoreNLP is an integrated suite of natural language processing tools including the part-of-speech (POS) tagger, the named entity recogniser (NER), the parser, the coreference resolution module, the sentiment analysis system.

From the set of tools offered by the Stanford CoreNLP toolkit, the sentiment analysis system is the module of our concern. This module implements a semantic compositional model for sentence sentiment classification [9]. The classification system is based on a recursive neural model to classify phrases and eventually, by recursion, to classify sentences. The training and evaluation of the compositional model makes use of the Stanford Sentiment Treebank corpus which contains labels for every syntactically phrase corresponding to thousands of sentences. Such corpus is based on the dataset of movie reviews collected by Pang and Lee [10].

Another important feature of the sentiment classifier is the number of sentiment classes for predicting the intensity of emotion in a sentence. To capture most of the intensity variation, the system implements five sentiment classes: very negative, negative, neutral, positive and very positive.

In short, this pre-processing task leads to the representation of a document (i.e. a review) as a collection of sentences and their corresponding polarities. This is the document representation that plays an important role for the processing of our composition model.

4.3   Model Parameters

Our composition model in (1) proposes a method which weighs the contribution of the sentiment classes C differently. In our particular case, since we make use of the sentiment classifier module of the Stanford CoreNLP toolkit, we estimate weights for five sentiment classes. To this end, the weights have been adjusted to a small held-out sentiment class and the weights vary in steps of 5 %.

The basic idea behind the setting of the weights is the heuristic intuition that the distribution of the sentences polarities determines the impact of the sentiment classes in a particular domain: a sentiment class which is prevalent among many sentences is not a good discriminator, and should be given less weight than one which occurs in few sentences. We show in Table 2 the weights for the best performing of our model where each row represents the weights corresponding to each sentiment class for a particular domain.

Table 2. Weights corresponding to each sentiment class for a particular domain

Domain	Very positive	Positive	Neutral	Negative	Very negative
Books	0.70	0.55	0.00	−0.20	−0.55
Compu	0.70	0.65	0.00	−0.20	−0.40
Movies	0.80	0.60	0.00	−0.30	−0.40
Music	0.70	0.55	0.00	−0.25	−0.45

Table 3 shows the distribution of the sentences polarities corresponding to each domain. As we can see, most of the sentences for each domain have been classified as *negative* and *very negative*. Thus, the weight assigned to these classes is smaller than the assigned to the *positive* and *very positive* classes. Now, within

Table 3. Distribution of the sentences polarities corresponding to each domain

Domain	Very positive	Positive	Neutral	Negative	Very negative
Books	23	332	293	903	38
Compu	37	501	453	1623	55
Movies	57	436	283	975	45
Music	55	652	538	1398	77

a particular polarity (i.e. positive), most of the sentences for each domain have been classified as *positive* rather than *very positive*. Thus, the weight assigned to the *positive* sentiment class is smaller than the assigned to the *very positive* class.

5   Results and Analysis

At first, we consider important to mention we are aware that the use of Machine Learning techniques in the implementation of a sentiment classifier outperform the results to be shown in this section. Training a classifier on a set of contextual features is the basic idea in the use of these techniques [11]. However, our approach is different: our purpose is to study whether a semantic composition model is a plausible alternative to categorise subjective texts.

As a baseline assessment, we first consider the intuitive mechanisms mentioned in the description of our composition model (Sect. 3). Table 4 displays the results obtained by using measures of central tendency as well as the results obtained with our model. Taking into account that the baseline accuracy for the different domains is 50 %, the results obtained by using the mode, the most frequent sentiment class in the set of sentences of each review, are poor. The total of negative review is recognised but this option struggles to recognise positive reviews.

Table 4. Precision results

Domain	Mode	Mean	Composition
Books	0.56	0.88	0.88
Compu	0.50	0.80	0.84
Movies	0.58	0.76	0.80
Music	0.54	0.86	0.90

On the other hand, the use of the mean produced better results, provided that a threshold value be assigned for optimisation purposes.[1] The average sentiment

---

[1] Since the sentiment classes are represented by a rank from 0 to 4, the value to be assigned to the threshold must be given in this rank. The best threshold values in our experimentation were: Books (2.60), Computers (2.45), Movies (2.65) and Music (2.75).

corresponding to the set of sentences of each review, improved the distribution of classification between positive and negative reviews. However, as we can see in Table 4, the best results were obtained in 3 out of 4 domains by our semantic composition model. Thus, the implementation of the idea that a sentiment class which is prevalent among many sentences should be given less weight than one which occurs in few sentences seems to be a plausible heuristic intuition. In fact, the propagation of the sentiment guided by the impact of non-neutral sentences also shows the plausibility of the composition function represented by the weighted additive model.

The distribution of the classification between positive and negative reviews is displayed by making use of a ROC graph. Three different classification results: mode, mean and semantic composition for each domain have been estimated. Figures 1 and 2 show the results corresponding to the Books and Music domains respectively.

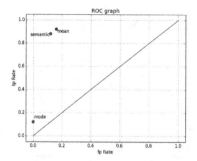 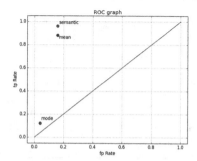

Fig. 1. Books results   **Fig. 2.** Music results

5.1   Analysis

In order to provide additional support to our model we extended our experimentation to the analysis of the dataset collected by Blitzer et al. [12]. This dataset is an interesting collection of product reviews corresponding to domains such as Books, DVDs, Music and Videos. There are 100 opinions per domain (50 per polarity), giving a total of 400 reviews in the collection. Table 5 shows the total number of sentences corresponding to each domain as well as the average number of sentences per text review. When we compare Table 5 with the information provided by Table 1, a huge difference can be noticed between the two collections. For example, the average number of sentences in a Book's review of the Blitzer corpus (8 sentences per review) is eight times smaller than the average number of sentences in the corresponding domain of the Taboada corpus (32 sentences per review).

By considering the two common domains between the two collections: Books and Music, the distribution of the classification between positive and negative reviews is also displayed by making use of a ROC graph. We can see in Figs. 3 and 4 how the best results in the Blitzer collection have been obtained by

Table 5. Number of sentences per domain (Blitzer corpus)

Domain	Sentences	Sentences/Review
Books	808	8
DVDs	915	9
Music	646	6
Videos	727	7

using our model. Moreover, when we compare Fig. 1 with Fig. 3, and Fig. 2 with Fig. 4, the results show considerable difference. In fact, the results corresponding to the Taboada collection (Figs. 1 and 2) exhibit better performance than the Blitzer collection. Does the average number of sentences per text review affect the results? Further investigation about this possible correlation is an interesting point to be explored.

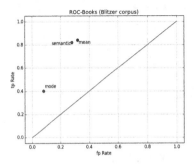

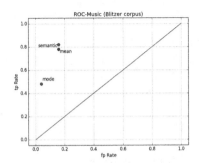

Fig. 3. Books results   **Fig. 4.** Music results

Another interesting point to investigate is to look for an alternative way of setting the weights of the C sentiment classes. Xia and Chai propose an algorithm to improve the common weighting scheme in Information Retrieval: tf-idf [13]. They claim *term frequency* is not the only discriminator to be considered in a term weighting scheme. Basically, they propose the concept of *term distribution* to analyse how uniform is the distribution of a linguistic term in both a single document and a document collection. In our particular case, it would be interesting to analyse the distribution of a sentiment class across a text review and a set of reviews corresponding to a particular domain.

Finally, to include machine learning techniques in the optimisation of the model's weights is also worth of investigation. In his work about semantic compositionality concerned with phrases, Guevara proposes the use of supervised learning for the optimisation of the weights by extracting an additional vector that denotes the target linguistic expression [7]. Exploring the adoption of a machine learning technique such as the *multi-layer perceptron* algorithm seems to be worth of exploration for the optimisation of our model's weights.

6 Conclusions and Future Work

In this paper, we focus our attention in the categorisation of a text review by implementing a semantic composition method. Specifically, we investigate semantic composition from the perspective of sentential sentiment composition: if the meaning of a sentence is a function of the meanings of its parts, the polarity of a sentence is a function of the polarities of its parts. In this way, we calculate in a systematic way the polarity of the review as a function of the impact of its subconstituents: each sentence and its contribution to the final emotion orientation of the review. The experimentation conducted with two different datasets showed how our approach is a plausible alternative to categorise subjective texts.

As part of our future work, we mentioned in the analysis section (Sect. 5.1) some perspectives to be investigated. To analyse how uniform is the distribution of the sentiment classes is an option [13], whereas to explore the incorporation of machine learning techniques is another alternative for the optimisation of the model's weights [7].

References

1. Partee, B., Meulen, A., Wall, R.: Mathematical Methods in Linguistics. Kluwer, Dordrecht (1990)
2. Moilanen, K., Pulman, S.: Sentiment composition. In: Proceedings of Recent Advances in Natural Language Processing (2007)
3. Manning, C., Surdeanu, M., Bauer, J., Finkel, J., Bethard, S., McClosky, D.: The Stanford CoreNLP natural language processing toolkit. In: Proceedings of 52nd Annual Meeting of the Association for Computational Linguistics: System Demonstrations, pp. 55–60 (2014)
4. Harris, Z.: Mathematical Structures of Language. Wiley, New York (1968)
5. Mitchell, J., Lapata, M.: Vector-based models of semantic composition. In: Proceedings of the 46th Annual Meeting of the ACL, Columbus, OH, pp. 236–244 (2008)
6. Mitchell, J., Lapata, M.: Composition in distributional models of semantics. Cogn. Sci. **34**, 1388–1429 (2010)
7. Guevara, E.: Computing semantic compositionality in distributional semantics. In: Proceedings of the Ninth International Conference on Computational Semantics, pp. 135–144 (2011)
8. Taboada, M., Anthony, C., Voll, K.: Creating semantic orientation dictionaries, pp. 427–432 (2006)
9. Socher, R., Perelygin, A., Wu, J., Chuang, J., Manning, C., Ng, A., Potts, C.: Recursive deep models for semantic compositionality over a sentiment treebank. In: Proceedings of EMNLP (2013)
10. Pang, B., Lee, L.: Seeing stars: exploiting class relationships for sentiment categorization with respect to rating scales. In: Proceedings of ACL, pp. 115–124 (2005)
11. Wilson, T., Wiebe, J., Hoffmann, P.: Recognizing contextual polarity in phrase-level sentiment analysis. In: Proceedings of HLT/EMNLP 2005 (2005)
12. Blitzer, J., Dredze, M., Pereira, F.: Biographies, bollywood, boom-boxes and blenders: domain adaptation for sentiment classification (2007)
13. Xia, T., Chai, Y.: An improvement to TF-IDF: term distribution based term weight algorithm. J. Softw. **6**(3), 413–420 (2011)

An Approach to Sentiment Analysis for Mobile Speech Applications

David Griol[1(✉)], José Manuel Molina[1], Araceli Sanchis[1], and Zoraida Callejas[2]

[1] Computer Science Department, Carlos III University of Madrid, Leganes, Spain
{david.griol,josemanuel.molina,araceli.sanchis}@uc3m.es
[2] Department of Languages and Computer Systems,
University of Granada, Granada, Spain
zoraida@ugr.es

Abstract. The integration of Sentiment Analysis and spoken conversational interfaces provides mutual benefits that enable using context-awareness information to enhance the performance of these interfaces, achieving a more efficient and proactive human-machine communication that can be dynamically adapted to the user's emotional state. In this paper, we describe a novel Sentiment Analysis approach combining a lexicon-based model for specifying the set of emotions and a statistical methodology to identify the most relevant topics in the document that are the targets of the sentiments. Our proposal also includes an heuristic learning method that allows improving the initial knowledge considering the users' feedback. We have integrated the proposed Sentiment Analysis approach into an Android-based mobile App that automatically assigns sentiments to pictures taking into account the description provided by the users.

Keywords: Sentiment Analysis · Emotion · Speech interaction · Mobile devices · Android

1 Introduction

Paralanguage information in spoken communication involves a number of intentional and not-intentional aspects, including the emotion of a speaker, the pragmatic force behind an utterance, the personality and demographic and cultural information about a speaker, etc.

Emotion is something intrinsically human and, as such, is part of our everyday communication. Sentiment Analysis (SA), also known as Opinion Mining, has the main objective of extracting subjective emotional information from a natural language source [10,16,20]. Basic Sentiment Analysis algorithms are focused on classifying the input according to its polarity towards a specific topic (positive, negative, or neutral) [15]. There are also advanced approaches that add an additional level of granularity by further identifying private states, such as opinions, emotions, sentiments, evaluations, beliefs, or speculations [23].

© Springer International Publishing Switzerland 2016
J.F. Quesada et al. (Eds.): FETLT 2015, LNAI 9577, pp. 96–107, 2016.
DOI: 10.1007/978-3-319-33500-1_9

Main applications of this field of study are currently related to marketing and social networks. Marketing applications are focused on determining customers' attitude towards products, which provides a very valuable information for companies to estimate products acceptance and market trends, offer products adapted to customers' requirements, and focus innovation on most demanded features. The growing importance of Sentiment Analysis coincides with the growth of social media for sharing thoughts about trending topics in reviews, forum discussions, blogs, micro-blogs, Twitter, and social networks [18].

In this paper, we describe a new proposal for Sentiment Analysis aimed to identifying the sentiments in the users' spoken utterances, instead of only detecting the positive or negative polarity. The set of defined sentiments has been extracted from Plutchik's wheel of emotions [17], which defines eight basic bipolar sentiments and another eight advanced emotions composed of two basic ones.

Our proposal for SA combines a lexicon-based model for specifying the relationships between terms and emotions, and a statistical approach to identify the most relevant topics in the document that contribute the different sentiments. With this combination, we overcome the disadvantages of simple Bag-of-words models [15], which do not differentiate between parts of speech (POS). Furthermore, our proposal includes an heuristic learning method that allows improving the initial knowledge in the model by considering the users' feedback.

We have used our proposal for SA to develop an Android-based mobile application for emotion detection from photographs, which has been designed based on important studies about key aspects of algorithmic inferencing of emotions that natural images arouse in people [5,6]. The developed App acts as a social network in which the users can share their photographs, know the emotions assigned to their descriptions, and compare how people from all around the world express their emotions about them.

2   Modeling the User Emotional State

As described in the previous section, emotions can affect the explicit message conveyed during the interaction with a spoken dialog system and also the actions that the user chooses to communicate with the system. According to [24], emotions can be understood more widely as a manipulation of the range of interaction affordances available to each counterpart in a conversation. They have also been recently considered as a very important factor of influence in decision making processes.

Three main classification levels have been defined for Sentiment Analysis: document-level, sentence-level, and aspect-level SA. Document-level SA aims to classify an opinion document as expressing a positive or negative opinion or sentiment. It considers the whole document a basic information unit (talking about one topic). Sentence-level SA aims to classify sentiment expressed in each sentence. Aspect-level SA aims to classify the sentiment with respect to the specific aspects of entities.

Sentiment Classification techniques can be roughly divided into machine learning approaches, lexicon based approaches, and hybrid approaches [11]. Machine Learning approaches apply this kind of algorithms and uses linguistic features. Lexicon-based approaches rely on a sentiment lexicon, a collection of known and precompiled sentiment terms. It is divided into dictionary-based approach and corpus-based approach which use statistical or semantic methods to find sentiment polarity. Hybrid approaches combine both approaches and is very common with sentiment lexicons playing a key role in the majority of methods.

Sentiment analysis is sometimes considered as a Natural Language Processing task for discovering opinions about an entity; and because there is some ambiguity about the difference between opinion, sentiment and emotion, they defined opinion as a transitional concept that reflects attitude towards an entity. The sentiment reflects feeling or emotion while emotion reflects attitude.

Within the framework of spoken conversational systems, emotions affect the explicit message conveyed during the interaction. They change people voices, facial expressions, gestures, and speech speed; a phenomenon addressed as emotional coloring [1, 13]. This effect can be of great importance for the interpretation of the user input. Emotions can also affect the actions that the user chooses to communicate with the system. According to [24], emotion can be understood more widely as a manipulation of the range of interaction affordances available to each counterpart in a conversation.

Despite its benefits, the recognition of emotions in dialog systems presents important challenges which are still unresolved. The first challenging issue is that the way a certain emotion is expressed generally depends on the speakers, their culture and environment [3]. Most work has focused on monolingual emotion classification, making an assumption there is no cultural difference among speakers. However, the task of multi-lingual classification has also been investigated [8].

Another problem is that some emotional states are long-term (e.g. sadness), while others are transient and do not last for more than a few minutes. As a consequence, it is not clear which emotion the automatic emotion recognizer will detect: the long-term emotion or the transient one. Thus, it is not trivial to select the categories being analyzed and classified by an automatic emotion recognizer. Linguists have defined extensive inventories of daily emotional states. A typical set is given by Schubiger [19] and O'Connor and Arnold [14], which contains 300 emotional states. However, how to classify such a large number of emotions, or even if it is tractable or practical, remains an open research question.

Also there is not a clear agreement about which speech features are most powerful in distinguishing between emotions. The acoustic variability introduced by the existence of different sentences, speakers, speaking styles, and speaking rates adds another obstacle because these properties directly affect most of the common extracted speech features such as pitch, and energy contours [2].

Related to these problems, some corpus developers prefer the number of utterances for each emotion to be almost the same in order to properly evaluate the classification accuracy. While balanced utterances are useful for controlled scientific analysis and experiments, they may reduce the validity of the data. For this reason, many other researchers prefer that the distribution of the emotions in the database reflects their real-world frequency [12]. In this case, the number of neutral utterances should be the largest in the emotional speech corpus. In addition, the recorded utterances in most emotional speech databases are not produced in the conversational domain of the system [9]. Therefore, utterances may lack some naturalness since it is believed that most emotions are out comes of our response to different situations.

In our proposal, we count only with the acoustic channel, so we carry out a prosody processing procedure like in multimodal systems such as SmartKom [22], but additionally consider other information sources related to the analysis of the words in the transcription in order to obtain better recognition rates (as we cannot rely on other modalities).

3 Our Approach for Sentiment Analysis of Spoken Utterances

The proposed model for Sentiment Analysis aims to extend common sentiment classification of text, which is usually focused on polarity, to a higher level so that the input texts are categorized by the emotions they evoke. Thus, the main goal is to recognize a specific set of human emotions instead of only detecting whether a piece of text is negative, neutral or positive. To do this, a limited set of emotions must be selected from one of the existing emotion classifications accepted by psychologist community.

After a detailed study of the principal affective models and considering computational requirements, we have selected a modification of the Hourglass emotion representation [4]. This model is based on Plutchik's wheel of emotions, which proposes eight basic emotions contrary to Ekman's initial classification that defines only six primary affection states. Although having more categories increases analysis complexity, Plutchik's model can be reduced into four categories -as there are four pairs of opposite emotions- so that, indeed, the analysis can be considered to turn out simpler. The proposed model is based on four key components.

The Knowledge Base (KB) contains the main information sources used by the Analysis Module to extract sentiment values from words. The Analysis Module completes the words analysis. By splitting texts in sentences an tokenizing words, this module can query the Knowledge Base to extract emotional information or know whether words are modifiers or carry an associated negation. Moreover, this module identify entities in the input text and track the number of occurrences of each one of them in a similar way bag-of-words models do this using occurrences vectors.

Once the entities have been identified and words are annotated with values from the KB, the Scoring Module computes the overall relevance of the entities and assigns a weighting factor for each of the words carrying emotional information, which are also known as concepts. A weight for each of the four independent emotional categories is then computed to classify the input text.

The last stage of the model deals with knowledge learning. To do this, the Learning Module takes as input the provided analysis from users when they disagree with the results of the Sentiment Analysis, and computes a learning factor to modify sentiment values of involved concepts.

3.1 Feature Extraction

The first step for emotion recognition is feature extraction. The aim is to compute features from the speech input which can be relevant for the detection of emotion in the users' voice. We extracted the most representative selection from the list of 60 features shown in Table 1. The feature selection process is carried out from a corpus of dialogs on demand, so that when new dialogs are available, the selection algorithms can be executed again and the list of representative features can be updated. The features are selected by majority voting of a forward selection algorithm, a genetic search, and a ranking filter using the default values of their respective parameters provided by the Weka toolkit.

The second step of the emotion recognition process is feature normalization, with which the features extracted in the previous phase are normalized around the user neutral speaking style. This enables us to make more representative classifications, as it might happen that a user 'A' always speaks very fast and

Table 1. Features defined for emotion detection from the acoustic signal [7, 12, 21]

Groups	Features	Physiological changes related to emotion.
Pitch	Minimum value, maximum value, mean, median, standard deviation, value in the first voiced segment, value in the last voiced segment, correlation coefficient, slope, and error of the linear regression.	Tension of the vocal folds and the sub glottal air pressure.
First two formant frequencies and their bandwidths	Minimum value, maximum value, range, mean, median, standard deviation and value in the first and last voiced segments.	Vocal tract resonances.
Energy	Minimum value, maximum value, mean, median, standard deviation, value in the first voiced segment, value in the last voiced segment, correlation, slope, and error of the energy linear regression.	Vocal effort, arousal of emotions.
Rhythm	Speech rate, duration of voiced segments, duration of unvoiced segments, duration of longest voiced segment and number of unvoiced segments.	Duration and stress conditions.

loudly, while a user 'B' always speaks in a very relaxed way. Then, some acoustic features may be the same for 'A' neutral as for 'B' angry, which would make the automatic classification fail for one of the users if the features are not normalized.

3.2 Knowledge Base

As previously described, the Knowledge Base contains the main information sources used by the Analysis Module to extract sentiment values from words. In our proposal, this information has been classified into the following categories:

– **Concepts:** A concept refers to the emotions associated to a specific pair of ($word - PoS$), where PoS (part of Speech) denotes the grammatical function of a word inside a predicate. Only the primitive form of a word is considered and the rest of derivative words take the same set of emotional values. The different categories of words are:
  - **Nouns:** Only the singular form is considered, although they may have an irregular plural that could be harder to identify. Nouns containing prefixes and suffixes are the only exception to this rule.
  - **Adjectives:** The positive form is considered and both comparative and superlative forms are discarded.
  - **Verbs:** The infinitive form is considered. Some exceptions are made for -ing forms acting as a noun (e.g., "The professor's reading about macro-economics was brilliant')'.
  - **Adverbs:** Only the positive form is considered, discarding comparative and superlative forms.
– **Modifiers:** Modifiers are denoted by an n-gram without associated sentiment states, which can increase, decrease or reverse the emotions of the associated concepts. They can be divided into two different categories:
  - **Intensity Modifiers:** This category is composed by those modifiers than may increase or decrease emotions expressed by concepts (e.g., "as much" or "a bit").
  - **Negators:** These modifiers reverse the global emotion associated to a concept (e.g., "not" or "never").

The NRC[1] and SenticNet[2] emotion lexicons have been used to complete the KB. Both are publicly available semantic resources for concept-level Sentiment Analysis. A total of 12,297 concepts are currently stored in the KB.

3.3 Parser Module

The parsing process of a sentence generates its semantically analysis containing part-of-speech tags organized in a tree of predicates. Between the set of general-purpose libraries currently available, we have selected OpenNLP[3]. This library

---

[1] http://www.saifmohammad.com/WebPages/lexicons.html.

[2] http://sentic.net/.

[3] https://opennlp.apache.org/.

supports the most common NLP tasks, such as tokenization, sentence segmentation, part-of-speech tagging, named entity extraction, chunking, parsing, and coreference resolution.

OpenNLP uses the Penn Treebank notation[4], which consider 36 sort of part of speech defined on the basis of their syntactic distribution rather than their semantic function. As a consequence nouns used in the function of modifiers are tagged as nouns instead of adjectives. Before parsing a text, it should be split into sentences by using the OpenNLP probabilistic *Sentence Detector*, which offers a precision of 94 % and a 90 % recall.

3.4   Emotion Classification Model

As stated before, our proposal uses an emotion representation model based on a modified version of the Hourglass model. The four independent categories that are considered for Sentiment Analysis consists of the following possible labels, described from negative maximum to positive maximum intensities, left to right:

- **Sensitivity:** [terror, fear, apprehension, neutral, annoyance, anger, rage]
- **Aptitude:** [amazement, surprise, distraction, neutral, interest, anticipation, vigilance]
- **Attention:** [grief, sadness, pensiveness, neutral, serenity, joy, ecstasy]
- **Pleasantness:** [loathing, disgust, boredom, neutral, acceptance, trust, admiration]

The algorithm applied for computing sentiment values of concepts is based both on distances of concepts to the selected representative nodes of all four categories and the weights assigned to each of them. The weights that are associated to each of the terms representing emotional categories are not trivial. They correspond to the maximum values of the different intensity levels of the original Hourglass model. The approach followed by the designed algorithm is to maximize the emotional intensities of concepts. Therefore, instead of using the returned distances of all the nodes of each category, only the most significant are considered. Figure 1 shows the designed algorithm.

3.5   Text Scoring Scheme and Adaptive Learning

Once the parsing process has finished and all the concepts, modifiers and negators have been properly tagged, it is possible to begin with the computation of the sentiment values of the text. The scoring process follows a bottom-up approach based on a fixed algorithm that relies on the Knowledge Base accuracy, a proximity based approach for modifiers, and a topic detection module to detect the most relevant topics of a text.

The method used to recognize most significant topic in the text follows a similar approach to a bag-of-words algorithm. During the parsing process the

---

[4] http://www.cis.upenn.edu/~treebank/.

Require: Term *concept*, Category *category*.
Ensure: Sentiment value for the specified category of the input term
1: $finalValue \leftarrow 0$
2: $maxDistance1 \leftarrow 0$
3: $maxDistance2 \leftarrow 0$
4: $weight1 \leftarrow 0$
5: $weight2 \leftarrow 0$
6: $auxiliaryMaxValue \leftarrow 0$ {Will store the max allowed value based on weights of maximum distances}
7: $targetNodes \leftarrow$ Nodes of the passed *category* {E.g: loathing, disgust, boredom, acceptance, trust, admiration for *Pleasantness*}
8: $distances \leftarrow$ Distances to $targetNodes$ {Array of distances preserving target nodes order}
9: $maxDistance2 \leftarrow 0$
10: **for all** $distances$ **do**
11:   $nodeWeight \leftarrow$ Weight associated to node whose distance is being considered
12:   **if** $nodeDistance > maxDistance1$ **then**
13:     $maxDistance1 \leftarrow nodeDistance$
14:     $weight1 \leftarrow nodeWeight$
15:   **else if** $nodeDistance = maxDistance1$ **then**
16:     **if** $|nodeWeight| = |weight1|$ **and** $(nodeWeight - weight1) \neq 0$ **then**
17:       $maxDistance2 \leftarrow nodeDistance$
18:       $weight2 \leftarrow MAX(|nodeWeight|, |weight2|)$ with corresponding sign
19:     **else if** $(nodeWeight - weight1) \neq 0$ **then**
20:       $maxDistance2 \leftarrow maxDistance1$
21:       $weight2 \leftarrow weight1$
22:       $weight1 \leftarrow MAX(|nodeWeight|, |weight1|)$ with corresponding sign
23:     **end if**
24:   **end if**
25: **end for**
26: **return** $finalValue$

Fig. 1. Computing sentiment value for an affection category of a concept

possible targets are counted so that a vector of entities and their number of occurrences is created. In order not to alter topics relevance, both plural and singular form of same entities share a common counter so that they are counted together. However, the hardest task within this approach is analyzing pronouns. Language ambiguity turns pronoun-to-noun mapping into a extremely difficult process.

The way sentences are weighted is based on entities occurrences. The process starts with smaller predicates going up to the bigger ones, from leaves to root. This bottom-up approach is needed as sentiments of a predicate are dependent of the smaller predicates contained on it. Therefore, the first step lies in resolving leaf predicates so algorithm can keep on computing bigger predicates and so on until the root sentiments are calculated, standing the root node for the whole user's utterance.

Let w_i be the weight of a predicate and n the total number of sibling predicates that are being combined, the sentiment value of a category for weighted predicates can be defined as:

$$S_w = \frac{\sum_{i=0}^{n} w_i * s_i}{\sum_{i=0}^{n} w_i}, \quad \begin{array}{l} \forall w_i > 0 \\ \forall s_i \neq 0 \\ s_i \in [-1, +1] \\ i = [0, n] \end{array} \tag{1}$$

Our proposal also integrates an adaptive learning process for improving the Knowledge Base used for Sentiment Analysis. This process uses Eq. 2 to consider the difference between the Sentiment Analysis output proposed by the SA algorithm and the feedback provided by the user. Let U be the set of sentiments of a text corrected by the user, M be the sentiments calculated by the SA algorithm, W_{C_s} be the weight of concept C for sentiment s, and A_c be the number of accumulated adjustments of concept C. Therefore the new value of each sentiment s for a concept C is defined as:

$$C_s = C_s + \frac{(U_s - M_s) * W_{C_s}}{1 + (A_C/1000)}. \tag{2}$$

4 A Mobile Application to Assess the Emotional Content of Photographs

The visual component of our proposal is an Android-based mobile application for Android OS consisting of a social network for sharing photographs. The minimum Android version required to run the App is Android 4.1 Jelly Bean, which is currently supported by more than 70 % of mobile phones.

Regarding the main use cases of the application, we can distinguish three main processes: accessing the App, browsing the gallery, and posting images. Accessing the application requires registering and using the login information stored in the mobile phone.

Browsing images can be done through two different screens. The initial screen shows all pictures shared in the system in chronological order starting from the newest one. A second screen allows users to filter the gallery by the sentiments identified after the analysis of their description.

The main functionality of the application is related to the posting process. This process includes the steps since the user decides to capture a photograph and share it on the App, to the assignation of sentiments considering the provided image description. The process can be divided in four stages: take photograph, type description, sentiment detection, and sentiment correction.

The first stage makes use of the default camera service client provided by Android. In the description screen, the user orally provides a description of the feelings or situations leading to the photograph which is about to be posted in a external server. Once the image is posted, the server starts running the implemented SA algorithm over the provided description. As a result of the analysis, the sentiments detected are shown with their respective intensities (anger versus fear, sadness versus joy, disgust versus trust, and surprise versus anticipation).

To finish the posting process, users have two possibilities: either accept the results of the analysis if they match user's real emotions or correct if the analysis is not accurate enough. If user chooses to click on accept button the post is finished and user is driven to main screen. However, in case the user's choice correspond to correcting the results, the App allows to select the right intensities for the sentiments expressed in the post description. After that, user just has to tap over the accept button to send the new values to the server so that the previous assigned intensities are substituted by the corrected ones and the post is updated.

The choice made by the user in this stage of the posting process has a significant meaning for the learning of the developed algorithm. On the one hand, whenever the user accept the results generated by the analysis means that the sentiments has been correctly detected and thus, the SA has succeeded in detecting the description's emotions. On the other hand, every time a user disagree with the outcomes of the analysis and chooses to provide the right sentiments of the introduced description, the new sentiments are not only used to update the classification, but also to update the algorithm by adjusting concepts weights as explained in the previous section.

From the design point of view, one the main characteristic of the App is that it acts as a social network, in which all posted photographs are publicly available. By publicly sharing all the uploaded pictures, the gallery will be an opportunity for comparing how people from all around the world express their emotions through their photographs.

A preliminary evaluation of the application has been already completed with the participation of 33 recruited users. The questionnaire with the following questions was defined for the evaluation: $Q1$: On a scale from 1 to 5, how much experience with smartphones do you have?; $Q2$: On a scale from 1 to 5, how much experience with Android do you have?; $Q3$: How often do you use image-sharing social networks (e.g., Instagram)?; $Q4$: On a scale from 1 to 5, how understandable the steps required in the different functionalities of the App are?; $Q5$: On a scale from 1 to 5, how accurate the detected emotions are?; $Q6$: Was it easy to use the App? The users were also asked to rate the system from 0 (minimum) to 10 (maximum) and there was an additional open question to write comments or remarks.

The results of the questionnaire are summarized in Table 2. As can be observed from the responses to the questionnaire, most of the users participating in the survey use smartphones and the Android OS, and not all of them usually access image-sharing social networks. Few of them had a previous knowledge about Sentiment Analysis.

Despite most of the participants agree with the set of sentiments used for representing emotions, almost half of them would prefer to have a larger list of sentiments available in the tool. Regarding accuracy, most of the users agreed that the overall performance of the analysis were from 3 to 5. Most of the users

Table 2. Results of the evaluation of the App by recruited users

	Min/max	Average	Std. deviation
Q1	4/5	4.17	1.19
Q2	3/5	3.06	1.47
Q3	2/5	2.83	1.25
Q4	4/5	4.17	0.68
Q5	3/5	4.63	0.45
Q6	4/5	4.83	0.37

found the application easy to use. The satisfaction with technical aspects of the application was also high, as well as the perceived potential to recommend its use. The global rate for the system was 8.6 (in the scale from 0 to 10).

5   Conclusions

In this paper, we have presented a novel Sentiment Analysis approach that combines a lexicon-based model for specifying the set of emotions and a statistical methodology to identify the most relevant topics in the document that are the targets of the sentiments. Our proposal for SA also includes an heuristic learning method that allows improving the initial knowledge considering the users' feedback. By means of our proposal, we overcome the main disadvantages of Bag-of-words models, which do not differentiate between parts of speech and usually lead to overweight most frequently used words. In addition, our proposal includes an heuristic learning method that allows improving the algorithm by updating the Knowledge Base.

We have used the proposed Sentiment Analysis approach to develop an Android-based mobile App that automatically assigns sentiments to pictures taking into account the description provided by the users. As future work, we want to extend the preliminary evaluation of the application to improve the proposed SA algorithm and carry out a comparative assessment with other SA algorithms. With the results of these activities, we will optimize the system, and make the application available in Google Play.

Acknowledgements. This work was supported in part by Projects MINECO TEC2012-37832-C02-01, CICYT TEC2011-28626-C02-02, CAM CONTEXTS (S2009/TIC-1485).

References

1. Acosta, J., Ward, N.: Responding to user emotional state by adding emotional coloring to utterances. In: Proceedings of the Interspeech 2009, pp. 1587–1590 (2009)

2. Banse, R., Scherer, K.: Acoustic profiles in vocal emotion expression. J. Pers. Soc. Psychol. **70**(3), 614–636 (1996)
3. Boehner, K., DePaula, R., Dourish, P., Sengers, P.: How emotion is made and measured. J. Hum. Comput. Stud. **65**, 275–291 (2007)
4. Cambria, E., Livingstone, A., Hussain, A.: The hourglass of emotions. In: Esposito, A., Esposito, A.M., Vinciarelli, A., Hoffmann, R., Müller, V.C. (eds.) Cognitive Behavioural Systems 2011. LNCS, vol. 7403, pp. 144–157. Springer, Heidelberg (2012)
5. Chen, C.-H., Weng, M.-F., Jeng, S.-K., Chuang, Y.-Y.: Emotion-based music visualization using photos. In: Satoh, S., Nack, F., Etoh, M. (eds.) MMM 2008. LNCS, vol. 4903, pp. 358–368. Springer, Heidelberg (2008)
6. Datta, R., Li, J., Wang, J.: Algorithmic inferencing of aesthetics and emotion in natural images: an exposition. In: Proceedings of the ICIP, pp. 105–108 (2008)
7. Hansen, J.: Analysis and compensation of speech under stress and noise for environmental robustness in speech recognition. Speech Commun. **20**(2), 151–170 (1996)
8. Hozjan, V., Kacic, Z.: Context-independent multilingual emotion recognition from speech signal. J. Speech Technol. **6**, 311–320 (2003)
9. Lee, C., Narayanan, S.: Toward detecting emotions in spoken dialogs. IEEE Trans. Speech Audio Process. **13**(2), 293–303 (2005)
10. Liu, B.: Sentiment Analysis and Opinion Mining. A Comprehensive Introduction and Survey. Morgan and Claypool Publishers, San Rafael (2012)
11. Medhat, W., Hassan, A., Korashy, H.: Sentiment analysis algorithms and applications: a survey. Ain Shams Eng. J. **5**(4), 1093–1113 (2014)
12. Morrison, D., Wang, R., DeSilva, L.: Ensemble methods for spoken emotion recognition in call-centres. Speech Commun. **49**(2), 98–112 (2007)
13. Murray, G., Carenini, G.: Detecting subjectivity in multiparty speech. In: Proceedings of the Interspeech (2009)
14. O'Connor, G., Arnold, J.: Intonation in Colloquial English. Longman, London (1973)
15. Pang, B., Lee, L., Vaithyanathan, S.: Thumbs up?: sentiment classification using machine learning techniques. In: Proceedings of the CEMNLP, pp. 79–86 (2002)
16. Pérez-Rosas, V., Mihalcea, R.: Sentiment analysis of online spoken reviews. In: Proceedings of the Interspeech (2013)
17. Plutchik, R., Kellerman, H.: Emotion: Theory, Research and Experience. Theories of Emotion, vol. 1. Academic Press, New York (1980)
18. Saif, H., He, Y., Fernandez, M., Alani, H.: Contextual semantics for sentiment analysis of twitter. Inf. Process. Manag. **52**(1), 5–19 (2015)
19. Schubiger, M.: English Intonation: Its Form and Function. Niemeyer Verlag, Halle (1958)
20. Serrano-Guerrero, J., Olivas, J., Romero, F., Herrera-Viedma, E.: Sentiment analysis: a review and comparative analysis of web services. Inf. Sci. **311**, 18–38 (2015)
21. Ververidis, D., Kotropoulos, C.: Emotional speech recognition: resources, features and methods. Speech Commun. **48**, 1162–1181 (2006)
22. Wahlster, W. (ed.): SmartKom: Foundations of Multimodal Dialogue Systems. Springer, Heidelberg (2006)
23. Wiebe, J., Riloff, E.: Creating subjective and objective sentence classifiers from unannotated texts. In: Gelbukh, A. (ed.) CICLing 2005. LNCS, vol. 3406, pp. 486–497. Springer, Heidelberg (2005)
24. Wilks, Y., Catizone, R., Worgan, S., Turunen, M.: Some background on dialogue management and conversational speech for dialogue systems. Comput. Speech Lang. **25**(2), 128–139 (2011)

Semantics for Social Media

Rodolfo Delmonte[(✉)] and Rocco Tripodi

Department of Language Science and Department of Computer Science,
Ca' Foscari University, Ca' Bembo, dd. 1075, 30123 Venice, Italy
{delmont,rocco.tripodi}@unive.it

Abstract. In this paper we present four experiments on the analysis Italian social media texts using a linguistically-based semantic approach. The experiments are respectively: two on newspaper articles about two political crises, one on a twitter corpus centered on political themes, and one on a case study of strategic plan programs of candidates to the presidency of our university. The analyses carried out by the same system, focus on semantic features of texts highlighting three main traits: "factivity" or factuality, "subjectivity" and polarity. The system uses semantic knowledge derived from deep linguistic analysis at propositional level to classify texts at a fine-grained level. As will be shown in the paper, linguistically-based semantic information allows for neat distinction of writing styles when comparing newspapers writing styles, for irony detection in tweets, and in different degrees, for making readability judgements.

1 Introduction

In this paper we present four experiments on Italian social media texts with the idea of an underlying unified approach which is strongly based on semantics, in particular propositional level and compositional semantic analysis. The experiments have all undergone a thorough evaluation and make use of a symbolic system for deep linguistic analysis called ITGetaruns [3]. Experiments carried out concern respectively: newspaper articles about political crises – two experiments dealing with 2011 and 2013 political Italian government crisis; one on a twitter corpus centered on political themes which was part of an international challenge and also had a task dedicated to irony detection; and a case study of strategic plan programs of candidates to the presidency of our university, where we compared writing styles and readability parameters again centered however on semantic deep analysis rather than simply using bag of words quantitative measurements. Semantic features of texts highlighted hing around clause level analysis and adopt a compositional level paradigm which allows for correct choices in sentiment analysis whenever there are polarity conflicts within the same sentence of even simply the same phrase. We have discovered that there are more relevant and less relevant semantic attributes that may contribute to distinguish texts from one another, and we have come up with three main traits: "factivity" or factuality, polarity and "subjectivity". In other words, semantic knowledge derived from deep linguistic analysis is used to tell factual from

© Springer International Publishing Switzerland 2016
J.F. Quesada et al. (Eds.): FETLT 2015, LNAI 9577, pp. 108–126, 2016.
DOI: 10.1007/978-3-319-33500-1_10

non-factual texts, and subjective from objective descriptions contained in every sentence – more on this distinction below. This is done in addition to other more popular features which highlight topics of discussion and polarity sentiment orientation. Semantic information allows for neat distinctions of writing styles but also of political orientation when comparing newspapers writing styles, for irony detection in tweets, and in different degrees, for making readability judgements. The system thus realized is partially comparable to a pipeline of modules which are also freely available on the web. However, we assume that in order to properly capture subjectivity and factuality expressed in a text or dialog any system needs a linguistic text processing approach that aims at producing semantically viable representation at propositional level. In particular, the idea that the task may be solved by the use of Information Retrieval tools like Bag of Words Approaches (BOWs) is insufficient. BOWs approaches are sometimes also camouflaged by a keyword based Ontology matching and Concept search [9], based on Senti-WordNet (*Sentiment Analysis and Opinion Mining with WordNet*) [2] – more on this resource below – by simply stemming a text and using content words to match its entries and produce some result [12]. Any search based on keywords and BOWs is fatally flawed by the impossibility to cope with such fundamental issues as the following ones, which Polanyi and Zaenen [22] named contextual valence shifters:

- presence of negation at different levels of syntactic constituency;
- presence of lexicalized negation in the verb or in adverbs;
- presence of conditional, counterfactual subordinators;
- double negations with copulative verbs;
- presence of modals and other modality operators.

It is important to remember that both Pointwise Mutual Information (PMI) and Latent Semantic Analysis (LSA) [15, 16] systematically omit function or stop words from their classification set of words and only consider content words. In order to cope with these linguistic elements, we propose to apply compositional semantic analysis at propositional level starting directly from a syntactic constituency or chunk-based representation. We implemented these additions in our system, thus trying to come as close as possible to the configuration which has been used for semantic evaluation purposes in challenges like Recognizing Textual Entailment (RTE) and other semantically heavy tasks [1,4,6]. The output of the system is an xml representation where each sentence of a text or dialog is a list of attribute-value pairs. In order to produce this output, the system makes use of a flat syntactic structure which however is based on subcategorization information to produce predicate-argument relations. Compositional semantic analysis is applied to verb complex at propositional level, thus splitting sentences into clauses. The output is a vector of semantic attributes associated to each clause and is memorized to be used for comparison between clauses. An important outcome is the distinction operated on the semantic content of each proposition into two separate categories: objective vs. subjective. This is obtained by searching for factivity markers again at propositional level [14]. In particular we take into account the following markers: modality operators such as intensifiers and diminishers, modal verbs, modifiers and attributes adjuncts at sentence level, lexical type of the verb (from ItalWordNet classification, and our own), subjects person (if 3rd or not), and so on. As will become clear below, we are using a lexicon-based [8, 25, 30] rather than a classifier-based approach, i.e. we make a fully supervised analysis where semantic features are manually associated to lemma and concept of the domain by creating a lexicon out of frequency lists. In this way the semantically labelled lexicon is produced in an empirical manner and fits perfectly the classification needs. This was needed in particular after we realized that available lexica where totally insufficient to cover the domain of political discourse. Of course we are aware of the intrinsic deficiencies of any such approach whenever irony, humour and figurative language is the target to be discovered, but see [15, 16] on the topic. The paper is structured as follows. Section 2 describes the lower level modules of the system; Sect. 3 describes the experiment on multi-dimensional political discourse analysis. Section 4 describes the higher level of the system and presents the experiment on twitters data; Sect. 5 presents work on text readability and persuasion related features; finally a conclusion.

2   Description of the System

The system called ITGetaruns shares its backbone with the companion English system which has been used – and documented – for a number of international challenges on Semantic and Pragmatic computing in English texts. It is organized around a manually checked subcategorized lexicon, a sequence of rules organized according to theoretical linguistics criteria and combines data-driven (bottom-up) and grammar-driven (top-down) techniques. Technically speaking, it is based on a shallow parser which in turn is based on a chunker and a NER and multiword recognizer. On top of this parser, there is constituent or phrase structure parser which sketches sentence structure. This is then passed to a deep dependency parser which combines constituent level information, lexical information, and a Deep Island Parser. The aim of this third parser is that of producing semantically viable Predicate-Argument Structures. Finally, on top of this level of representation, the Semantic and Pragmatic System is built. Conceptually speaking, the deep island parser (hence DIP) is very simple to define, but hard to implement. A semantic island is made up by a set of A/As (Argument/Adjuncts) which are dependent on a verb complex (hence VCX). In Italian, Arguments and Adjuncts may occur in any order and in any position: before or after the verb complex, or be simply empty or null. Their existence is determined by constituents surrounding the VCX. The VCX itself can be composed of all main and minor constituents occuring with the verb and contributing to characterize its semantics. We are here referring to: proclitcs, negation and other adverbials, modals, restructuring verbs (lasciare/let, fare/make, etc.), and all auxiliaries. Tensed morphology can then appear on the main lexical verb or on the auxiliary/modal/restructuring verb. Gender can appear on the past participle when the verb takes auxiliary ESSERE, or when a complement is duplicated by Clitic Left Dislocation – this is typical of Italian. The DIP is preceded by a tagger which is accompanied by a multiword expression labeler. Tagged input is passed to an augmented context-free parser that works on top of a chunker. The chunker collects main constituents on the basis of a Recursive Transition Network of Italian and then passes the output to a cascaded sentence level parser. Constituents are labeled with usual grammatical relations on the basis of syntactic subcategorization contained in our verb lexicon of Italian counting some 17,000 entries. There are some 270 syntactic classes which differentiates also the most common prepositions associated to oblique arguments. Linear position and precedence in the input string is assumed at first as a valid criterion for distinguishing SUBJects from OBJects. Adjustments are executed by the semantic parser, which is responsible for the final relabeling of the output. The DIP receives the output of the surface parser, a list of Referring Expressions and a list of VCX. Referring expressions are all nominal heads accompanied by semantic class information collected in a previous recursive run through the list of the now lemmatized and morphologically analyzed input sentence. It also receives the output of the context-free parser. The DIP searches for SUBJects at first and assumes it is positioned before the verb and close to it. In case there is none such chunk available, the search is widened if intermediate chunks are detected: they can be Prepositional Phrases, Adverbials or simply Parentheticals. If this search fails, the DIP looks for OBJects adjacent to the verb and possibly separated by some intermediate chunk. They will be relabeled as Subjects. Conditions on the A/As boundaries are formulated in these terms: between current VCX and prospective argument there cannot be any other VCX. Additional constraints regard presence of relative or complement clauses which are detected from the output chunked structure. The prospective argument is deleted from the list of Referring Expressions and the same happens with the VCX. The same applies for the OBJect, OBJect1 and OBLique. When arguments are completed, the parser searches recursively for ADJuncts which are PPs or AdverbialPs, using the same boundary constraint formulation above. Special provisions are given to copulative constructions which can often be reversed in Italian: the predicate coming first and then the subject NP. The choice is governed by looking at referring attributes, which include definiteness, quantification, distinction between proper/common noun. It assigns the most referring nominal to the SUBJect and the less referring nominal to the predicate. In this phase, whenever a SUBJect is not found from available referring expressions, it is created as little_pro and morphological features are added from the ones belonging to the verb complex. After updating Referring Expressions with Grammatical Relations, the parser searches the most adequate Semantic Role to be associated to it. This is again taken from a lexicon of corresponding verb predicates and works according to the type of overall Predicate-Argument Structure (hence PAS). The SUBJect is in fact strictly depending on the semantics associated to the verb, but in case of ambiguity the system delays the assignment of semantic role until a complete PAS is obtained. In this phase, passive diathesis is checked in order to apply a lexical rule from LFG [7], that assigns OBJect semantic role to the SUBJect of the corresponding passive form of the verb predicate. The PAS thus obtained, is then enriched by a second part of the algorithm which adds empty or null elements to untensed clauses. The system starts from little_pros and looks for local possible antecedents. An additional semantic function is activated in this phase of analysis and is the creation of verbal multiwords, constituted by the concatenation of a verb lemma and the head of its object, as for instance "tener conto"/take_into_account, which transforms the main predicate TENER into TENER_CONTO. In this operation, the system has available a list of light verbs of Italian which are the most frequent main component of the compound. Then the OBJect complement head is extracted and the concatenation is searched in a specialized dictionary of verbal multiwords of Italian. The OBJect is then erased from the list of arguments and the Argument/Adjunct distinction is updated according to the new governing predicate.

3   Print Press Discourse

For the elaboration of preliminary conclusions on the process of the change of the Italian government and president of government, we collected, stored and processed – partially manually, partially automatically – relevant texts published by three national on-line newspapers having similar profiles[1] (Table 1).

Table 1. Quantitative data of print press experiment

```
• Selection of articles by keywords (Berlusconi & Monti) and by relevance
• 250 articles x newspaper = 750 articles
• 150K tokens x newspaper = 415K tokens
• OMBB = 9987 sentences x 199003 tokens
• OMAB = 8792 sentences x 185544 tokens
• PTMB = 1494 sentences x 30029 tokens
   - Total sentences = 20273
   - Total tokens = 415000
```

Here above general numerical data of the database of news articles collected for 2011 Italian political crisis, where OMBB stands for One Month Before Berlusconi resigned, OMAB for After Berlusconi resigned and PTMB, Period of Time intervening after Berlusconi resigned and Mario Monti nomination as new Prime Minister. In the graph below we represent semantic data related to Polarity, Factivity, Subjectivity and Diathesis for the three newspapers investigated and for the three time spans. Peculiarities of this graph representation is the peak of Corriere's nonfactual and subjective index in the OMAB period, which is preceded by a similar movement in the PTMB interval. In the graph data are projected as differences from the mean and may thus show up or down the zero line.

For this reason we wanted to see what happened in the following similar event which took place last year in 2013 when Mario Monti was obliged to

---

[1] www.corriere.it, www.liberoquotidiano.it, www.repubblica.it.

resign and the government was entrusted to a new Prime Minister. We decided
to evaluate manually the data produced by our system and this was the topic
of a Master thesis which was also checked personally by myself. The experiment
setup required a smaller amount of data to be checked manually and a clear
indication of choices to be made when annotating different types of modality.
Instructions to the annotator were as follows:

- check tensed factive propositions from untensed ones
- check tensed propositions were modality is present as one or double feature
  and compute them as nonfactive
- check factive gerundives and participles from infinitivals
- check simple infinitivals from factive past or complex infinitivals
- check propositions dependent from a nonfactive matrix clause
- check for lexically triggered subjectivity – semantically marked verb classes.

Computation time on a tower MacPro equipped with 6 Gb RAM and 1 Xeon
quad-core was approximately 2 h (Fig. 2).

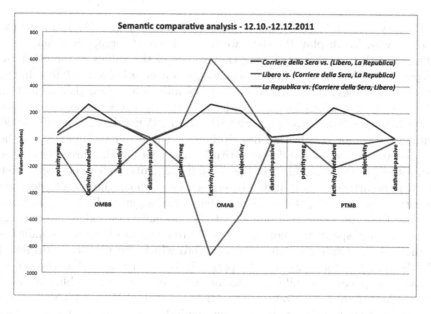

Fig. 1. Comparative semantic analysis of three Italian newspapers.

3.1 The Syntactic and Semantic Analysis

In Fig. 1, we present comparative semantic polarity and subjectivity analyses of
the texts extracted from the three Italian newspapers. On the graph we show
differences in values for four linguistic variables: they are measured as percent
value over the total number of semantic linguistic variables selected from the
overall analysis and distributed over three time periods on X axis.

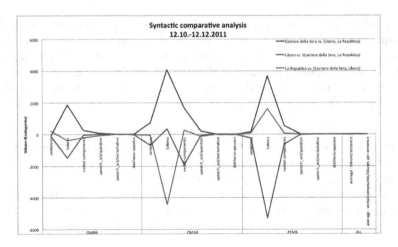

Fig. 2. Comparative quantitative and syntactic analysis of three Italian newspapers.

Polarity and subjectivity can only be measured in a relative and not in an absolute way. To display the data we use a simple difference formula, where Difference value is subtracted from the average of the values of the other two newpapers for that class. Differences may appear over or below the 0 line. In particular, values above the 0x axis mean they assume positive or higher values than below the 0x axis, which have a negative import. The classes chosen are respectively: 1. propositional level polarity with NEGATIVE value; 2. factivity which contains values for non factual descriptions; 3. subjectivity; 4. passive diathesis. We can now evaluate different attitudes and styles of the three newspapers with respect to the three historical periods: in particular we can now appreciate whether the articles report facts objectively, i.e. without the use of additional comments documenting the opinion of the journalist/journal. Or if it is rather the case that the subjective opinion of the journalist/journal is present only in certain time spans and not in others. Chronological difference is indicated by the three separate contiguous subsets into which the values are displayed, OMBB coming Before OMAB. The period in Between is placed at the end for its lower intermediate significance. So for instance, *Corriere*, the blue or darker line, has higher nonfactive values in two time spans, OMBB and PTMB; *Repubblica* values soar in OMAB. In the same period *Libero* has the lowest values; whereas in OMBB, *Libero* and *Corriere* have the highest values when compared with *Repubblica*. PTMB clearly shows up as a real intermediate period of turmoil which introduces a change: here *Repubblica* becomes more factual whereas *Libero* does the opposite. Subjectivity is distributed very much in the same way as factuality, in the three time periods even though with lesser intensity. *Libero* is the most factual newspaper, with the least number of subjective clauses. Similar conclusion can be drawn from the use of passive clauses, where we see again that *Libero* has the lowest number. The reasons for *Libero* having the lowest number of nonfactive clauses in OMAB, needs to be connected with the highest number of NEGATIVE polarity clauses, related to the nomination of Monti instead of Berlusconi, and is felt and is communicated to its readers as less reliable, less trustable, trustworthy. Uncertainty is clearly shown in the intermediate period, PTMB, where *Corriere* has again the highest number of nonfactual clauses.

We also saw above that Libero is the newspaper with the highest number of nonfactual and subjective clauses in the OMAB time period: if we now add this information to the one derived from the use of positive vs. negative words, we see that the dramatic change in the political situation is no longer shown by the presence of a strong affective vocabulary, but by a way of presenting important concepts related to the current political and economic situation, which becomes vague and less factual after Berlusconi resigned.

With one month before the Berlusconi's resignation (OMBB), we can highlight sentence structure of the three dailies as follows: *Il Corriere* has a sentence structure with a rich vocabulary (words, verbal component). There are on average 21 tokens and 3 verbal compounds per sentence; on average 25 % of all verbal compounds are subjective. Questions are 492, with only 1 exclamative. *Libero* has on average 20 tokens and 2.4 verbal compounds per sentence; on average only 20 % of all verbal compounds are subjective. Only 249 sentences are interrogative but 9 are exclamative sentences; more verbs are used with passive diathesis. *La Republica* has on average 18 tokens and 2.2 verbal compounds per sentence; on average only 25 % of all verbal compounds are subjective. Only 196 sentences are interrogative and 8 are exclamative sentences; only 22 verbs are used with passive diathesis. In other words, *Il Corriere* has longer sentences, more questions but less exclamatives than other newspapers. *Libero* has a style with less subjective verbal compounds, that is it uses more factive verbal compounds and structures, more below on this topic. *La Repubblica* uses shorter sentences than the other newspapers, remarkably less interrogatives.

The general quantitative data presented here in Table 2, are derived from the second experiment and show that a similar situation to the previous one took place. Here we intended to evaluate the output of the system against manual annotation. In fact, even though the dataset used was much smaller, only 6000 sentences compared to 20000 of the previous experiment, we can clearly see that overall, for *Il Corriere* the number of nonfactive and subjective propositions is higher than the ones of the other newpapers. It constitutes the 37 % against the 29 % of *Libero* and the 34 % of *La Repubblica*. Similar proportions can be found for Subjectivity, where *Il Corriere* has again 36 % against 30 % of *Libero* and 34 % of *La Repubblica*. These data are furthermore confirmed by the distribution of the two features in the sentences into which texts are organized: as can be noticed, *Repubblica* has the highest number of sentences, followed by *Libero* and then *Corriere*. In addition, propositions or clauses per sentence are much higher in *Corriere* than in the other newspapers, thus indicating a higher semantic density. Weighted data are shown in the graphs below. Results of the evaluation are shown in the graphs in Figs. 3 and 4. As can be gathered, mistakes in automatic annotation of semantic features is strongly related to error propagation in the pipeline that constitutes the system.

Table 2. Quantitative overall data of the experiment for subjectivity and nonfactivity evaluation

Newspapers	Tot. subject	Tot. nonfact	Errs. nonfact	Errs. subject	No. sents	No. propos. structs
Corriere	1377–37 %	2504–36 %	236	196	1804–31 %	5514–37 %
Libero	1142–29 %	1971–30 %	159	47	1965–34 %	4424–29 %
Repubblica	1290–34 %	2264–34 %	152	36	2042–35 %	5048–34 %
TOTALI	3809	6739	547	279	5811	14986

Thus mistakes in tagging and in dependency parsing may affect the final outcome. Additional errors are caused by problems in the semantic predicate-argument structure building process where in some cases verbs have been wrongly collapsed in one single Verb Complex even though they constituted separated items. However, error percentages for nonfactivity is overall at 8.2 %, while errors percentages for subjectivity is slightly lower, at 7.35 %. As can be noticed from Table 2, and graphs below, *Il Corriere* is by far the more difficult newspaper to analyse in terms of semantic features. The great majority of errors are present in *Il Corriere* which also has the highest number of propositions but not the lowest number of sentences. This fact means that sentences in *Il Corriere* are much longer and more complex to read.

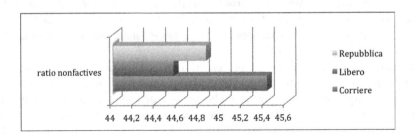

Fig. 3. Proportion of nonfactive propositions for the three newspapers

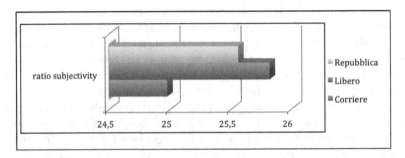

Fig. 4. Proportion of subjective propositions for the three newspapers

When compared to number of propositions we see a different distribution of data with *Il Corriere* having the highest number of nonfactive proposition but *Libero* having the highest number of Subjective propositions.

4   The System ITGetaruns and Irony Detection in Tweets

Sentiment Analysis is based on propositional level semantic processing, which in turn is made of two basic components: Predicate Argument Structures (hence PAS) and Verbal Complex (hence VCX) semantics. Semantic mapping is based on a number of intermediate semantic representations which include, beside diathesis:

- Discourse Domain; Change in the World; Subjectivity and Point of View; Speech Act; Factuality; Relevance; Polarity.

At first we compute (grammatical)Mood and Tense from the VCX which may contain auxiliaries, modals, clitics, negation and possibly adverbials in between. From Mood_Tense we derive a label that is the compound tense and this is then used together with Aspectual lexical properties of the main verb to compute Change_in_the_World. Basically this results into a subclassification of events into three subclasses: Static, Gradual, Culminating. From Change_in_the_World we compute (Point_of_)View, which can be either Internal (Extensional/Intensional) or External, where Internal is again produced from a semantic labeling of the subcategorized lexicon along the lines suggested in linguistic studies, where psych(ological) verbs are separated from movement verbs etc. Internal View then allows a labeling of the VCX as Subjective and otherwise, Objective. Eventually, we look for negation which can be produced by presence of a negative particle or be directly in the verb meaning as lexicalised negation. Negation, View and Semantic Class, together with presence or absence of Adverbial factual markers are then used to produce a Factuality labeling.

One important secondary effect that carries over from this local labeling, is a higher level propositional level ability to determine inferential links intervening between propositions. Whenever we detect possible dependencies between adjacent VCXs we check to see whether the preceding verb belongs to the class of implicatives. We are here referring to verbs such as "refuse, reject, hamper, prevent, hinder, etc." on the one side, and "manage, oblige, cause, provoke, etc." on the other (for a complete list see [33]). In the first case, the implication is that the action described in the complement clause is not factual, as for instance in "John refused to drive to Boston", from which we know that "John did not drive to Boston". In the second case, the opposite will apply, as in "John managed to drive to Boston".

Two notions have been highlighted in the literature on discourse: foreground and background. The foreground is that part of a discourse which provides the main information; in a narrative, for example, the foreground is the temporal sequence of events; foreground information, then, moves the story forward.

The background, on the contrary, provides supportive information, such as elaborations, comments, etc., and does not move the story forward. To compute foreground and background information, three main rhetorical relations are assigned by the algorithm (for a deeper description see [5, 7]) in the form of attribute-value pairs, or features: Discourse Domain, Change in the World, Relevance.

The Discourse Domain of a sentence may be "subjective", indicating that the event or state takes place in the mind of the participant argument of the predicate and not necessarily in the external world. Else it may be "objective", indicating that the action described by the verb affects the whole environment. A sentence may also describe a "change in the world", in case we pass from the description of one situation to the description of another situation which precedes or follows the former in time but which is not temporally equivalent to it; we have then the following inventory of changes: null (i.e. no change), gradual, culminated, earlier, negated. The third value, the Relevance of a sentence, corresponds to the distinction between foreground and background which has been discussed above.

We have now to explain the way each utterance receives its set of values: the algorithm relies heavily on grammatical cues, i.e. those linguistic elements encoded in the grammar of a language which allow interpretation without the intervention of pragmatic or non-linguistic elements such as conversational implicatures, presupposition or inferencing. The cues we make use of are chiefly extracted from the verb and are semantic category, polarity, tense, aspect. The procedure is very simple from a theoretical point of view: once the algorithm has recognized a cue, it assigns a value to the sentence. Note that we distinguish between direct and indirect speech portions of the text, since the perspective is not the same in the two cases.

- DISCOURSE DOMAIN: to assign the point of view of a sentence, the algorithm checks the sem(antic)_cat(egory) of the main verb of the sentence and a number of other opacity operators, like the presence of future tense, a question or an exclamative, the presence of modals, etc.
- CHANGE IN THE WORLD: to establish whether a clause describes a change or not, and which type of change it describes, the algorithm takes into account four parameters: polarity (i.e. affirmative or negative), domain, tense and aspect of the main verb.

If polarity is set to NO (i.e. if the clause is negative), CHANGE is negated; but if the verb describes a state, CHANGE is null because a stative verb can never express a change, apart from the fact that it is affirmed or negated. Thus, if DISCOURSE DOMAIN is subjective and the verb is stative, CHANGE is null: this captures the fact that, in such a case, the action affects only the subject's mind and has no effects on the outside world. In all other cases the algorithm takes into account tense and aspect of the main verb and obeys the following rules: if tense is simple present, CHANGE is null; if tense is simple past, CHANGE is culminated; if tense is pluperfect, CHANGE is earlier; if tense is the "imperfetto" – this tense belongs to Italian but not to English verb system – and describes a state, CHANGE is null, but if it describes an activity, a process, an accomplishment, or if it is a mental activity, CHANGE is gradual.

- FACTIVITY: this relation may only assume two values: factive and nonfactive. A factive relation is assigned every time Change is non Null. Other sources of information may be used to trigger factivity, and that is the presence of a factive predicate, like a presuppositional verb, "know".

We now turn to the cues for direct speech. Once the algorithm has recognized a clause to be in direct speech, the CLAUSE TYPE value is dir_speech/prop. The DISCOURSE DOMAIN is also subjective: this is so because direct speech reports the thoughts and perceptions of the characters in the story, so that any intervention of the writer is left out. As far as CHANGE is concerned, the algorithm obeys the following rules: if the main verb is in the *imperative* mood, CHANGE is null because, although the imperative is used to express commands, there is no certainty that once a command has been imparted it is going to be carried out. If verb is in the indicative mood, and it is in the *future*, CHANGE is null as well since the action has still to take place; if we have a past tense such as the *perfect* or *pluperfect*, CHANGE is culminated or earlier, respectively; if tense is *present*, the algorithm checks its aspect: if the verb describes a state, CHANGE is null, otherwise (i.e. if the verb describes an activity) CHANGE is gradual. Finally, negative and positive polarity is carefully weighted in case the sentence has a complex structure, and is compositionally driven, taking care of cases of double negations. Positives are so marked when the words searched in the input sentence belong to the class of socalled "Absolute Positives", i.e. words that can only take on positive evaluative meaning. The same applies for Negative polarity words, when they belong to a list of "Absolute Negatives", like swear words.

4.1   Results and Discussion

The automatic detection of irony in texts or tweets is a highly difficult task [26,27]. In the task organized by EVALITA 2014 more than 20 groups participated, however only seven presented results. The baseline for the irony task was 0.4441 and best result was 0.5759. Here below is the table of our results for the three tasks of Sentipolc (see [2]).

In Table 3 we report percent values of our system performance[2]. In a final column we registered our placement in the graded scale of final results. As can be

Table 3. Results of ITGetaruns for all tasks.

Task	F-ScoreTot	Prec0	Rec0	F-score0	Prec1	Rec1	F-score1	Rank
Subjectivity	52.24	34.79	30.26	32.37	75.71	68.83	72.11	9th/9
Polarity pos	51.81	72.97	*81.58*	77.03	43.13	16.05	23.39	10th/11
Polarity neg	51.81	60.97	*77.00*	68.05	62.03	28.19	38.77	10th/11
Irony	49.29	88.29	77.54	82.57	15.66	16.39	16.02	4th/7

---

[2] Final overall results are available online, http://www.slideshare.net/vivianapatti9/evalita-sentipolc14.

noticed, best result has been achieved for irony detection. In general, we can note the following: in our experiment, there has always been an attempt to favour Recall rather than Precision, and also an attempt to reduce False Positives. This is testified by a better scoring in those values associated to Prec0, Rec0 and F-score0 where 0(zero) refers to the separate evaluation carried out on assignment of 0 value to irony (i.e. no irony) for a given tweet. As can be noticed, both Polarity and Irony have by far better scoring in 0 s than in 1 s. On the contrary, Subjectivity has much better scores in 1 s than in 0 s. We assume that this is due to annotation criteria which don't match our linguistic rules. We marked with bold italics those scores that have better ranking individually, and both coincide with Recall0 in Polarity. Recall0 for Polarity Pos is 81.58, which corresponds to the 4th rank in the list of 12 (not considering the baseline); Recall0 for Polarity Neg is 77.00 which represents the best result of all systems. Going back to annotation criteria, one of our basic rule for Subjectivity matching is presence of 1st and 2nd person morphology in the main verb complex associated to the main or root clause. We noticed that this does not always coincide with annotations associated to the tweets.

As to irony, the starting point was subjectivity and non-factivity: the majority of ironic statement were in fact exclamatives or rhetorical questions. Then we implemented a number of additional features which have increased Precision quite significantly but somehow decreased Recall. One of these features was the possibility to highlight the use of alterations in Ironic tweets which are used to express "Exaggeration". The algorithm was based on our Morphological Analyzer that in turn is based on linguistic rules for alterations and a root lexicon of Italian made up of some 90,000 entries (see [14, 15]). We also used our classification of Emoticons, which however proved not to be a highly significant contribution in the overall evaluation.

5   Quantitative Stylostatistic and Semantic Analysis

In this section we present an evaluative predictive stylistic analysis of texts which constitute a balanced attempt at combining quantitative approaches with deep pragmatic, semantic and syntactic analysis. The analysis starts from the idea that a document style – in this case a strategic political program at university level – be analyzable quantitatively at word level, but also structurally and relationally at syntactic, semantic and pragmatic level, by looking at frequent usage of certain concepts and structures inspiring negativity or positivity, but also factuality and subjectivity. Texts analyzed were made freely available on a forum of candidates to the election of president or dean of the university. Here in Table 4 absolute values of texts analysed are listed.

Final evaluation was organized around the following parameters which contributed with positive or negative marks to create a graded final evaluation scale:

1. NullSubject - Positive: A higher quantity of null subjects indicates the intention to create a highly cohesive text, trying not to overcharge coreferring

Table 4. Absolute quantitative data

Candidate	Tokens	Types	Rare words	Sents	Little pro	Propos. structs
Bertinetti	4992	1561	1341	162	200	585
Brugiavini	2841	987	852	91	119	308
Bugliesi	13210	2483	1899	463	541	1232
Cardinaletti	5346	1479	1243	167	159	469
LiCalzi	14376	3120	2516	769	720	1624

mechanisms by the use of repeated slightly different linguistic forms and or descriptions indicating some property of the same entity.

2. Subjective Props - Negative: A higher number of subjective propositions indicates a tendency on the narrator to express one's ideas in a non objective manner.

3. Negative Props - Negative: A higher use of negative propositions, where there is a usage of negation and/or negative adverbials associated to the verb, a negative governing predicate of the whole sentence, a negatively marked predicative complement, or simply a negatively interpreted argument of the predicate, is computed as a stylistic trait which is not proactive but tends to be assertive by contrasting what has been affirmed by others.

4. Nonfactive Props - Negative: The use of non-factive propositions indicates a stylistic tendency to expose one's ideas using unreal tenses and moods – subjunctive, conditional, future and indefinite tenses – and in this way making no straightforward reference to real current objective facts.

5. Props/Sents - Negative: The ratio that indicates the number of propositions/clauses per sentence is considered having negative import to stress that a higher complexity at semantic level implies a worsening of the readability index.

6. Negative Ws - Negative: The number of negative words as a ratio of the total number of words used is also computed as a negative parameter.

7. Positive Ws - Positive: The contrary applies when the number of positive words used as a ration of total number of words is higher than the negative one.

8. Passive Diath - Negative: The number of passive diathesis used in the text is computed as a negative parameter because it obscures the agent of the action described by the sentence.

9. Token/Sents - Negative: Number of tokens per sentence is computed as a negative factor, again with reference to a possible increase in complexity.

10. Vr - Rw - Negative: This measure evaluates socalled vocabulary richness on the basis of the ration between RareWords – Hapax/Dis/Tri Legomena included in the Rank List – and the rest of the Types. The higher are RareWords the less Readable will be the text and the more complex the style.

11. Vr - Tt - Negative: Another quantitative evaluation this time based simply on Types and Tokens ratio.

Together with these general parameters, we deemed it necessary to analyse in some detail one concept which didn't figure at the same rank level in the Rank

List organized for the five candidates. It is the concept related to STAFF. This word is used as a collective noun in Italian as in English, to characterize the group of people working in the university organization. Obviously, they constituted by far the majority of the people voting, but internal regulations gave their vote a different weight from the one of the teaching staff. We excluded two candidates – Brugiavini and Bertinetti – from this computation due to the almost inexistence of the concept in their Rank List. Their final mark will not modify their position in the graded scale, however, seen that they where already in the same lower slots (Table 5).

Table 5. Usage of the concept *Staff*

	Noun	Adjective	Multiword	Total
LiCalzi	22	4	5	17
Cardin	11	2	4	7
Bugliesi	37	2	5	32

Texts however, also use alternative linguistic descriptions, usually more specific ones, to talk about Staff. So we updated the table above as indicated below.

Table 6. Usage of the concept *Staff* and relative values

	Noun	Adjective	Multiword	Hyponyms	Total
LiCalzi	22	4	5	7	0.32
Cardin	11	2	4	8	0.72
Bugliesi	37	2	5	6	0.16

In Table 6 we reported absolute values under Hyponyms but then relative values under Total, by computing ratios between Hyponyms and total number of referring expressions for Staff. Here below are the three main figures reporting all the relations indicated above. However we show final graded evaluation first. It is important to know that Bugliesi won the contest and became our new Rector or president of Ca' Foscari University. As can be easily noticed from the general Tables 7 and 8, almost all his parameters are evaluated to 5 or 4, excluding the one related to non-factive propositions and the final ones more domain related.

Table 7. Final graded scale on the basis of 11 parameters.

	1	2	3	4	5	6	7	8	9	10	11	Tot
Bugliesi	4	4	5	2	4	5	4	5	4	5	5	47
LiCalzi	5	1	2	5	5	2	1	2	5	4	4	36
Brugiavini	3	2	4	4	2	1	5	3	2	1	1	28
Cardinaletti	2	5	3	1	3	4	2	1	1	3	3	27
Bertinetti	1	3	1	3	1	3	3	4	3	2	2	27

Table 8. Final graded scale on the basis of 13 parameters.

	1	2	3	4	5	6	7	8	9	10	11	12	13	Tot
Bugliesi	4	4	5	2	4	5	4	5	4	5	5	3	3	53
LiCalzi	5	1	2	5	5	2	1	2	5	4	4	4	4	44
Cardinaletti	2	5	3	1	3	4	2	1	1	3	3	5	5	37
Brugiavini	3	2	4	4	2	1	5	3	2	1	1	2	1	31
Bertinetti	1	3	1	3	1	3	3	4	3	2	2	1	2	30

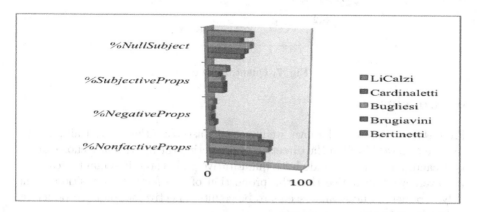

Fig. 5. Syntactic-semantic data.

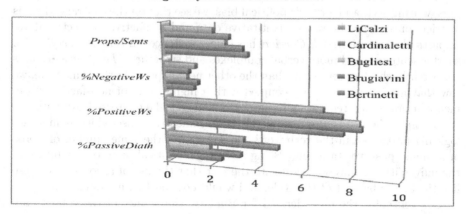

Fig. 6. Affective and semantic data.

Evaluation was done using 5 mark as the highest positive mark depending on the value of the parameter: for instance, when a negative parameter is evaluated, 5 assumes the value of the least negative case and 1 the most negative one. The opposite applies to positive parameters (Figs. 5, 6 and 7).

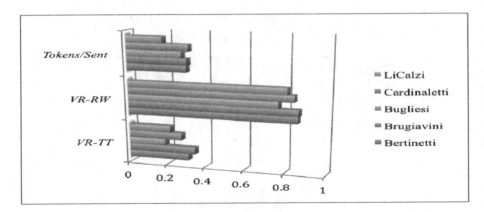

Fig. 7. Quantitative data

6 Conclusion

From the data presented above, we may easily ascertain that almost all parameters seem to work well in the direction of predicting the appropriate choice of the best candidate. They include both quantitative and syntactic-semantic features. However, we may notice that the proportion of non-factive proposition is the only parameter whose value seems to be counter-effective or counterproductive: in fact it ranks the winner fourth.

If we compare these results to the ones obtained when characterizing newspapers' writing style and/or their political bias, we see that on the contrary, there is a strict concordance between quantitative data and nonfactive/subjective data. In particular, we see that *Il Corriere* has longer sentences, *Libero* has a style with shorter sentences and more verbal complexes and structures. *La Repubblica* uses consistently shortest sentences than the other newspapers. At the same time, we saw that longer sentences are coupled with a higher level of non-factive statements because they require more untensed clauses to be expressed. All with the exception of the use of the future: this is an indicative tense which is morphologically integrated in the verb word and so requires the same number of words of a simple positive statement. Being a strategic political plan for the future of the university, we may well admit to the fact that the use of a nonfactive tense like the future be part of the style, and would not constitute a deviation.

So eventually, the data derived for Bugliesi, the winner of the contest, are fully justified on a more general scale: they are characterized by short sentences, more frequently used words, less clauses per sentence, less negative statements, less passive diatheses, more null subjects. All of these features confirm the general knowledge about text readability that less complex and shorter texts/sentences are better understood and may reach a wider audience than the opposite. There is a wide literature on the topics, we have drawn from [1, 17–19, 22, 25, 28, 29, 33].

We may thus conclude that semantic classification tasks at a fine-grained level as required in our three cases take advantage of deep semantic analysis at different degrees, however. In particular, text readability judgements require additional quantitative features related to text semantic complexity level, which are more syntactically based. These features are nonetheless strongly based on a deep linguistic analysis that predicts null subjects and clause level subdivision of sentences. For these reasons, we assume that a symbolic system, which is not tied to any specific training corpus and can thus be applied to any text may be more suitable than an automatic machine learning approach which is more domain dependent.

References

1. Bagga, A., Pustejovsky, J., Zadrozny, W. (eds.): Syntactic and Semantic Complexity in Natural Language Processing Systems: NAACL-ANLP 2000 Workshop. Association for Computational Linguistics, New Brunswick (2000)
2. Basile, V., Bolioli, A., Nissim, M., Patti, V., Rosso, P.: Overview of the Evalita 2014 SENTIment POLarity Classification Task. In: Proceedings of EVALITA 2014, Pisa (2014)
3. Bos, J., Delmonte, R. (eds.): Semantics in Text Processing (STEP), Research in Computational Semantics, vol. 1. College Publications, London (2008)
4. Esuli, A., Sebastiani, F.: SentiWordNet: a publicly available lexical resource for opinion mining. In: Proceedings of the 5th Conference on Language Resources and Evaluation LREC 2006 (2006)
5. Delmonte, R.: Computational Linguistic Text Processing - Logical Form, Logical Form, Semantic Interpretation, Discourse Relations and Question Answering. Nova Science Publishers, New York (2007)
6. Delmonte, R., Tonelli, S., Tripodi, R.: Semantic processing for text entailment with VENSES. In: TAC 2009 Proceedings Papers (2010). http://www.nist.gov/tac/publications/2009/papers.html
7. Delmonte, R.: Computational Linguistic Text Processing - Lexicon, Grammar, Parsing and Anaphora Resolution. Nova Science Publishers, New York (2009)
8. Delmonte, R.: Predicate argument structures for information extractionfrom dependency representations: null elements are missing, 2013. In: Lai, C., Giuliani, A., Semeraro, G. (eds.) DART 2012: Revised and Invited Papers. Studies in Computational Intelligence, vol. 515, pp. 25–49. Springer, Heidelberg (2012)
9. Delmonte, R., Gifu, D., Tripodi, R.: Opinion and factivity analysis of Italian political discourse. In: Basili, R., Sebastiani, F., Semeraro, G. (eds.) Proceedings of the 4th Italian Information Retrieval Workshop, IIR 2013, Pisa. CEUR Workshop Proceedings (CEUR-WS.org), vol. 964, pp. 88–99 (2013). http://ceur-ws.org
10. Delmonte, R., Pallotta, V.: Opinion mining and sentiment analysis need text understanding. Advances in Distributed Agent-based Retrieval Tools. Advances in Intelligent and Soft Computing, pp. 81–95. Springer, Heidelberg (2011)
11. Delmonte R., Pianta, E.: IMMORTALE - Analizzatore Morfologico, Tagger e Lemmatizzatore per l'Italiano. In: Atti V Convegno AI*IA "Cibernetica e Machine Learning", pp. 19–22. Napoli (1996)
12. Delmonte, R., Pianta, E.: Immortal: how to detect misspelled from unknown words. In: BULAG, PCUF, pp. 193–218. Besanon (1998)
13. Delmonte, R.: Italian lemmatization by rules with getaruns. In: Magnini, B., Cutugno, F., Falcone, M., Pianta, E. (eds.) EVALITA 2011. LNCS, vol. 7689, pp. 239–248. Springer, Heidelberg (2012)
14. DuBay, W.H.: The Principles of Readability. Impact Information, Costa Mesa (2004)

15. Heilman, M., Collins-Thompson, K., Eskenazi, M.: An analysis of statistical models and features for reading difficulty prediction. In: Proceedings of the 3rd Workshop on Innovative Use of NLP for Building Educational Applications. Columbus (2008)
16. Heilman, M., Collins-Thompson, K., Callan, J., Eskenazi, M.: Combining lexical and grammatical features to improve readability measures for first and second language texts. In: HLT-NAACL 2007, pp. 460–467. Association for Computational Linguistics, Rochester (2007)
17. Kim, S.-M., Hovy, E.: Determining the sentiment of opinions. In: Proceedings of the 20th International Conference on Computational Linguistics (COLING 2004), pp. 1367–1373 (2004)
18. Kao, J., Dan, J.: A computational analysis of style, affect, and imagery in contemporary poetry. In: NAACL Workshop on Computational Linguistics for Literature (2012)
19. Feng, L., Jansche, M., Huenerfauth, M., Elhadad, N.: A comparison of features for automatic readability assessment. In: Proceedings of the 23rd International Conference on Computational Linguistics (COLING 2010), Beijing (2010)
20. Pang, B., Lee, L.: A sentimental education: sentiment analysis using subjectivity summarization based on minimum cuts. In: Proceedings of the 42nd Annual Meeting of the Association for Computational Linguistics (ACL), pp. 271–278 (2004)
21. Pennebaker, J.W., Booth, R.J., Francis, M.E.: Linguistic Inquiry and Word Count (LIWC). http://www.liwc.net/
22. Petersen, S.E., Ostendorf, M.: A machine learning approach to reading level assessment. Comput. Speech Lang. **23**, 86–106 (2009)
23. Polanyi, L., Zaenen, A.: Contextual valence shifters. In: Wiebe, J. (ed.) Computing Attitude and Affect in Text: Theory and Applications, pp. 1–10. Springer, Dordrecht (2006)
24. Pollack, M., Pereira, F.: Incremental interpretation. Artif. Intell. **50**, 37–82 (1991)
25. Pollard, S., Biermann, A.W.: A measure of semantic complexity for natural language systems. In: Bagga et al. Proceedings of the 2000 NAACL-ANLP Workshop on Syntactic and semantic complexity in natural language processing systems, vol. 1, pp. 42–46 (2000)
26. Reyes, A., Rosso, P., Buscaldi, D.: From humor recognition to irony detection: the figurative language of social media. Data Knowl. Eng. **74**, 1–12 (2012)
27. Reyes, A., Rosso, P.: On the difficulty of automatically detecting irony: beyond a simple case of negation. Knowl. Inf. Syst. **40**(3), 595–614 (2013)
28. Roark, B., Mitchell, M., Hollingshead, K.: Syntactic complexity measures for detecting mild cognitive impairment. In: Proceedings of the Workshop on BioNLP 2007: Biological, Translational, and Clinical Language Processing, BioNLP 2007, pp. 1–8. Association for Computational Linguistics, Stroudsburg (2011)
29. Saur, R., Pustejovsky, J.: Are you sure that this happened? Assessing the factuality degree of events in text. Comput. Linguist. **38**(2), 261–299 (2012)
30. Taboada, M., Brooke, J., Tofiloski, M., Voll, K., Stede, M.: Lexicon-based methods for sentiment analysis. Comput. Linguist. **37**(2), 267–307 (2011)
31. Turney, P.D., Littman, M.L.: Inference of semantic orientation from association. ACM Trans. Inf. Syst. (TOIS) **21**(4), 315–346 (2003)
32. Tonelli, S., Manh, K.T., Pianta, E.: Making readability indices readable. In: Proceedings NAACL-HLT 2012 Workshop on Predicting and Improving Text Readability for target reader populations, pp. 40–48. Montral (2012)
33. Wiebe, J., Wilson, T., Cardie, C.: Annotating expressions of opinions and emotions in language. Lang. Resour. Eval. **39**(2), 165–210 (2005)

Semantic Similarity Reasoning

Luigi Di Caro[✉] and Guido Boella

University of Turin, Turin, Italy
{dicaro,boella}@di.unito.it

Abstract. The cornerstone of current Computational Linguistics research is the computation of semantic similarity between lexical items or some of their conceptualization in available semantic resources such as WordNet. However, measures for semantic similarity (and/or relatedness) usually work with numerical outputs, which are then used to solve tasks related to word disambiguation rather than information retrieval. In this paper, we start from the limitations of using numeric-based similarity measures, proposing a novel approach to provide *explanations* of similarity, even if still calculated through statistical (and thus numerical) analyses. This may allow a novel, fine-grained and context-based *similarity reasoning* over lexical entities. In this contribution, we define the concept of *semantic similarity reasoning* and a method of extraction from ConceptNet, a large common-sense resource. Finally, we present a number of hypotheses of how such shift of paradigm could represent a new building block of future natural language technologies.

Keywords: Ditributional semantics · Semantic similarity · Similarity reasoning

1 Introduction

The fundamental principle of the research on Computational Linguistics is the computation of similarity scores between texts at different levels of granularity: letters, words, sentences, paragraphs, and documents. For example, an automatic spell checker needs to calculate a distance between an input sequence of letters and an entry from a dictionary to propose correct forms, considering features such as the number of different or equal letters rather than the position of the letters on a keyboard to estimate possible typing errors, and so forth. On a bigger scale, since words represent symbolic entities which can be connected to multiple meanings, an important task (i.e., Word Sense Disambiguation, or WSD) is the selection of the right meaning which better fits with (and thus which is close to) the contextual information. For instance, the term *count* can mean *nobleman* or *sum*, but the sentence *The total count is lower than 10* clearly identifies one among the two choices. Generally speaking, *comparing things* is the key element every computational task is asked to solve in some way.

The work has been funded by the project *Semantic Burst: Embodying Semantic Resources in Vector Space Models*, financed by Compagnia di San Paolo.

J.F. Quesada et al. (Eds.): FETLT 2015, LNAI 9577, pp. 127–138, 2016.
DOI: 10.1007/978-3-319-33500-1_11

The concept of similarity has been extensively studied in the Cognitive Science community, since it is fundamental in the human cognition. We tend to rely on similarity to generate inferences and categorize objects into kinds when we do not know exactly what properties are relevant, or when we cannot easily separate an object into separate properties. When specific knowledge is available, then a generic assessment of similarity is less relevant [14]. By generalization, if one has complete knowledge about the reasons why objects have specific properties, similarity should be no longer relevant [16]. In the light of these thoughts, this work is inspired by what stated in [6], i.e., if we know about a property, then this knowledge is directly used to make inferences rather than a *one-size-fits-all* similarity [12].

There exist several annotated datasets related to similarity and relatedness between words, such as wordsim-353 [4] and SimLex-999 [7]. In general, a large part of the proposed computational systems aims at finding *relatedness* between words, instead of similarity. Relatedness is more general than similarity, since it refers to a generic correlation (like *cradle* and *baby*, that are words representing dissimilar concepts which, however, share similar contexts and are thus semantically related).

The key problem of the current way similarity (or relatedness) is considered is its numerical (and static) nature. As the authors of the above-mentioned datasets state in their works, the inter-annotation agreement is quite low (around 50–70%). The reason is trivial, however: people can give different degrees of importance to the existing characteristics of the concepts under comparison. If we ask one to say how much *dog* is similar to *cat*, the right answer can only be *"it depends on what you mean for similar"*. While we can all agree about the fact that the concept *dog* is quite related to *cat*, we cannot say 0.7 rather than 0.9 (in a [0,1] range) with certainty. Different aspects can be taken into account: are we measuring the form of the animal, or its behaviour? In both cases, it depends on which part of the animal and which actions we are considering to make a choice. For instance, dogs use to return thrown objects. From this point of view, dogs and cats are dissimilar.

Currently, state-of-the-art similarity approaches are not able to capture the dynamic aspect of word meaning; besides essential properties captured by the definition of a noun, it may assume role-specific meanings in particular contexts. For example *dog* can be replaced with *pet* in the context of *A dog is a good companion*, but not in *That dog is similar to a wolf*. This means that the words *dog* and *pet* are exchangeable in particular contexts only (i.e., when the dog plays the role of a pet). Words are what the linguist Ferdinand de Saussure called *signifiers*, i.e., empty boxes that link a symbol (the word) with a concept. Missing from the existing literature in computational approaches is Saussure's third element, i.e., the mental representation. How the semiotic triangle works is that a word (or phrase) is a symbol which brings to mind a mental representation of a real or imagined object with all its known characteristics and connotations. Humans do not merely reason on the lexical level, so there is the need to include semantics within the semantic similarity models and their output.

In this paper, we start from these motivations to propose a novel model of semantic similarity called *similarity reasoning* which relies on a distributional analysis over the knowledge retrieved from a semantic resource and associated to the words under comparison. However, in its yet preliminary stage, this contribution does not enter into the details of important questions such as the concrete/abstract nature of the concepts and the application on multiple languages. Nevertheless, it represents a look beyond the current vision of natural language technologies towards a statistical but more logical-based understanding of meaning and similarity.

2   Background and Related Work

As already mentioned in the introduction, research on Computational Linguistics is usually focused on the calculation of similarity scores between pieces of texts at different granularity (e.g., word, sentence, discourse, etc.) [11]. Although many measures have been proposed in the literature, this work aims at changing the current paradigm of the commonly-accepted notion of semantic similarity, and thus is mostly related to the cognitive and linguistic theories that orbit around the concept of *meaning* and from which is directly inspired. In the end, semantic similarity is about meanings ascribed to lexical entities, so this work can be also considered as a modest attempt to shift the current focus towards a more semantic level of analysis.

2.1   Meaning of Meaning

Over time, brands of philosophers, linguists, semioticians, cognitive psychologists and computer scientists, have investigated the nature of meaning and a number of different perspectives, hypotheses and theories have emerged (e.g., the semantic theories, the mentalist theories, the structuralist theories, the distributional semantics theory, etc.). The difficulty of defining the *meaning of meaning* has to do with tricky issues like lexical ambiguity and polysemy, vagueness, contextual variability of word meaning, and so forth. As a matter of fact, words are organized in lexicon as a complex network of semantic relation which are basically subsumed within Saussure's paradigmatic (the axis of combination) and syntagmatic (the axis of choice) axes [18].

2.2   Meaning as Relation

Some authors [2] have already suggested theoretical and empirical taxonomies of semantic relations consisting of some main families of relation (such as contrast, similars, class inclusion, part-whole, etc.). As Murphy points out [15], lexicon has become more central in linguistic theories and, even if there is no a widely accepted theory on its internal semantic structure and how lexical information are represented in it, the semantic relations among words are considered in scholarly literature as relevant to the structure of both lexical and conceptual information and it is generally believed that relations among words determine meaning.

2.3   Meaning as Interaction

Interaction is another important aspect that has been investigated in literature. Indeed, the actions change the type of perception of an object, which models itself to fit with the context of use. Then, the Gestalt theory [8] contains different notions about the perception of meaning according to interaction and context. In particular, the core of the model is the complementarity between the *figure* and the *ground*. In our case, a word is the figure and the ground is the context that lets emerge its specific sense. Finally, James Gibson introduced the concept of affordances as the cognitive cues that an object exposes to the external world, indicating ways of use [5]. In cognitive and computational linguistics, this theory can be inherited to model words as objects and contexts as their interaction with the world.

2.4   Meaning as Distribution

Distributional Analysis of natural language, such as Distributional (or vector-based) semantics, exploits Harris's distributional hypothesis (later summarized with Firth's sentence *you shall know a word by the company it keeps*) and sees a word meaning as a vector of numeric occurrences (i.e., frequencies) in a set of linguistic contexts (documents, syntactic dependences, etc.). This approach provides a semantics of similarity which relies on a geometrical representation of the word meanings, and so in terms of vector space models (VSMs, [17]). This view has been recently gaining a lot of interest and success, also due to the growing availability of large corpora from where to obtain statistically-significant lexical correlations. Data Mining (DM) techniques fully leveraging on VSMs and Latent Semantic Analysis (LSA) [3] have been successfully applied on text since many decades for information indexing and extraction tasks, using matrix decompositions such as Singular Value Decomposition (SVD) to reconstruct the latent structure behind the above-mentioned distributional hypothesis, often producing concept-like entities in the form of words clusters sharing similar contexts. However, distributional approaches are usually good in finding lexical relatedness rather than similarity. The IBM's Question Answering system called *Watson*[1] uses Big Data to find specific answers with amazing results, but in IBM itself, the frontier of CL has already moved on to finding explanations to statements by means of Argument Mining, e.g., why is a given chemical dangerous. For these latest developments, a semantics of similarity is no longer sufficient, and other aspects of meaning related to Formal Semantics are becoming increasingly important [10].

3   The Proposal

In this section, we present the details of the proposal. In particular, we firstly give an overview of the paradigm and the definition of *explanation* in the context

---

[1] http://www.ibm.com/smarterplanet/us/en/ibmwatson/.

of the proposed concept of *semantic reasoning*. Then, we briefly illustrate and motivate the used semantic resource (i.e., ConceptNet) and the similarity dataset on which we based our experimentation. Finally, we detail the mathematical model and some running examples.

3.1 Overview of the Approach

The proposed approach aims at giving a semantic explanation of the similarity between two lexical entities. Given two words w_1 and w_2, what is usually done by standard methods is to compare their *word semantic profiles*, i.e., contextual lexical item sets. For example, given the word *cat* and *dog*, their profiles can share terms such as *pet, fur, claws* and so forth. These word profiles are extracted from co-occurrences rather than available resources (using relations such as synonyms, hypernyms, meronyms, etc.). Instead, our idea is to replace *blind* lexical overlappings by a meaningful matching of semantic information.

Definition of Explanation. The first step is the extraction of individual *word explanations* for both w_1 and w_2. A word explanation is a relation-based model which correlates conceptual properties (i.e., semantic information related to physical aspects) to functional features (i.e., semantic information related to behavioural facts). This vision is directly inspired by studies in Cognitive Science, mainly derived by the concept of *affordances*, introduced by Gibson in 1977 [5]. Generally speaking, the functions of a world entity is latently communicated through its physical properties. In a sense, there is a strong relationship (or *interaction*) between a physical property and some function of the object. In our case, given a word w and a semantic relation r, we use a semantic resource KB to extract all the words that show the semantic instance $r - w$. For example, if $r = has$ and $w = fur$, the query would be $has - fur$, and the result will contain the set of words which have that semantic information in KB, for example $\{cat, bear, ...\}$. Then, we obtain all the semantics related to such retrieved terms, building a matrix which correlates the conceptual properties to the functional features (by using Pointwise Mutual Information, as later described). For example, the property of *having claws* can usually correlate with the fact of *climbing trees*.

The set of matrices $M = <M_1, M_2, ..., M_{|r|}>$ (one for each semantic relation provided by KB) represent the semantic explanation e of a single word w, i.e., all *property-to-function interactions* related to the semantics around a single word w (from different perspectives, by considering all types of semantic relations in KB). For instance, the analysis of the words *bear* and *cat* may lead to explanations that differently associate functionalities to the property of having claws (e.g., *climbing trees* for a cat, *killing people* for a bear). Anticipating the details of the next step, in case of a semantic comparison between the words *cat* and *bear*, the claws-property will be not used as an element of similarity, because of their different *meanings* in terms of enabled behaviour. This point represents a radical novelty with respect to state-of-the-art approaches which only considers lexical overlappings without taking into account their actual context-based meaning (Fig. 1).

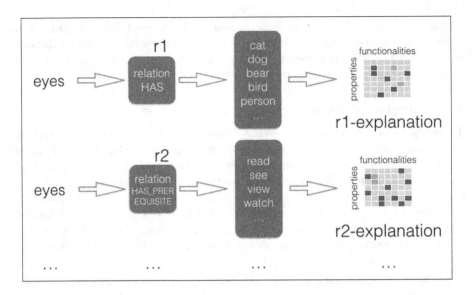

Fig. 1. Example of relation-based explanations for the word *eyes* (*cats*, *dogs*, etc. have eyes; *read, see*, etc. have the prerequisite of having eyes). The matrices represent the correlation between properties and functionalities of these resulting words.

Definition of Similarity Reasoning. For each relation r in KB, we then obtain the relative explanations for the words w_1 and w_2, namely e_1 and e_2 respectively. At this point, we make a comparison of e_1 and e_2 by aligning the vectors of the relative sets of matrices M_w1 and M_w2 (i.e., each vector of the matrices of e_1 is aligned to zero or some vector of the matrices of e_2 if they represent the same property). In case of a non-empty alignment (i.e., e_1 and e_2 share some identical property), the two property-vectors are compared in terms of their functional features. Again, in case the vectors share identical functionalities, the numeric product of their weight represents a score (and thus the importance) of one *similarity reasoning instance*.

In words, a similarity reasoning instance that links a row row_1 of one r_x-matrix in e_1 and a row row_2 of one matrix r_y in e_2 would mean that everything that is related with w_1 through the relation r_x has a property p with an overlapping functional distribution with the same property p of what is related with word w_2 through the relation r_y.

For example, if $w_1 = cat$, $w_2 = tree$, $r_x = partOf$ and $r_y = usedFor$, one similarity reasoning instance contains the property $p = claws$, since claws are parts of a cat and they are used for climbing. In a sense, this instance *explain* an aspect of similarity between cats and trees in terms of a property (the claws), saying that cats can climb the trees through this physical feature (Fig. 2).

Considering the entire set of relations r in KB, the similarity reasoning of two input words w_1 and w_2 is the r-based sets of similarity reasoning instances which represent the direct matching between identical properties and their functional aspects.

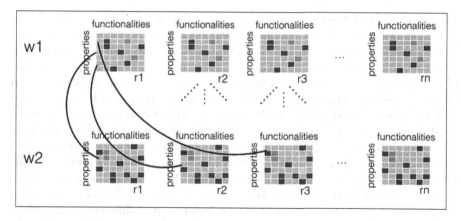

Fig. 2. Scheme of the similarity reasoning process, which correlates matrix rows (property vectors) of the explanations of the two words under comparison.

3.2 Semantic Resource: ConceptNet

This proposal needs a source of semantic information to associate to the input words. While linguistic resources such as WordNet [13] and VerbNet [19] constitute highly-precise (top-down) sources of information, they often cover very few semantic relations. Common-sense knowledge, on the other side, represents a much larger set of semantic features, which, however, is affected by noise, cultural differences and sparseness. The Open Mind Common Sense[2] project developed by MIT collected unstructured common-sense knowledge by asking people to contribute over the Web. In this paper, we started focusing on ConceptNet [20], that is a semantic graph that has been directly created from it. In contrast with the mentioned linguistic resources, ConceptNet contains common-sense facts which are also connected to behavioural information, so it perfectly fits with the proposed model. We used the following semantic relations as *properties*: partOf, madeOf, hasA, definedAs, hasProperty. Then, we used the following relations as *functionalities*: capableOf, usedFor, hasPrerequisite, motivatedByGoal, desires, causes.

3.3 Dataset and Humans Agreement

In this paper we used the dataset named *SimLex-999* [7], which contains one thousand word pairs that have been manually annotated with similarity scores. The inter-annotation agreement is 0.67 (Spearman correlation), highlighting the complexity of the task (and somehow underlining the motivations of this proposal). *SimLex-999* also includes word pairs of another dataset, *WordSim-353* [4], that contains a mix of relatedness- and similarity-based items.

3.4 Algorithm

In this section we present the details of the algorithm for the extraction of explanations and the final similarity reasoning between two input words.

---

[2] http://commons.media.mit.edu/.

The pseudo-code of the entire approach is shown in Algorithm 1, 2, and 3. Given two input words w_1 and w_2, and a semantic resource KB, the system creates a set of semantic reasoning instances, one for each type of relation r among the whole set R of relations in KB. Each semantic reasoning instance is the result of a comparison between the two explanations of the two words according to a specific target semantic relation. In detail, we query the semantic resource KB with the relation-word pair $r - w$. At this point, we query KB with each of these words, and collect the co-occurrence values between conceptual and functional information. In particular, we build a matrix M of $nr = |P|$ rows and $nc = |F|$ columns, where P and F are respectively the set of property features and the set of functional features, and where each value $M_{i,j}$ contains the co-occurrence of the property p_i with the functionality f_j (with $0 < i < |P|$, and $0 < j < |F|$). Once the matrix of co-occurrences M is calculated, it is then transformed in a PMI-based matrix where each value $M_{i,j}$ is replaced with:

$$M'_{i,j} = \frac{P_{i,j}}{P_i * P_j}$$

where $P_{i,j}$ is the probability of having a non-zero co-occurrence value for the property p_i and the functionality f_j (that is, $M_{i,j} > 0$) in the semantics of the input word, while P_i and P_j are the individual probability to find the property p_i and the functionality f_j respectively. The utility of M' is to capture the strength of the associations between properties and functionalities also considering their individual frequency. Each horizontal vector of a matrix M_r represents a word explanation, i.e., how a property is related to some functionalities with respect to the considered semantic information related to $r - w$. Finally, we align the explanations of w_1 with the ones of w_2. Given a property vector of M'_{w1}, if the property is also contained in M'_{w2}, we calculate their matching functionalities. If the matching is not empty, we add the semantic reasoning instance in the final result. At the end of the process, the similarity reasoning output of the two input words w_1 and w_2 will be the set of similarity reasoning instances obtained for all the relations R.

Data: word w_1, word w_2, semantic resource KB
Result: A set of similarity reasoning instances, given two input words
          and a semantic resource (ConceptNet in our case).
R = set of relations in KB;
result = empty set of relation-based similarity reasoning instances;
for *each relation r in R* **do**
 | Explanation e_1 = getExplanation(w_1, KB, r);
 | Explanation e_2 = getExplanation(w_2, KB, r);
 | SimilarityReasoning sr = getReasoning(e_1, e_2);
 | result.add(r, sr);
end
 return result;
Algorithm 1. Main method to return the final Similarity Reasoning output.

Data: word w, relation r, semantic resource KB
Result: A set of explanations for a given word.
$context_{words} = \text{query}(KB, \text{r}, w)$;
for *each word cw in $context_{words}$* **do**
    $cw_{semantics} = \text{query}(KB, cw)$;
    $cw_{conceptual} = \text{selectConceptualSemantics}(cw_{semantics})$;
    $cw_{functional} = \text{selectFunctionalSemantics}(cw_{semantics})$;
    $\text{updateCoOccurrence}(cw_{conceptual}, cw_{functional})$;
end
return the PMI values calculated over the co-occurrences;
Algorithm 2. Method *getExplanation* for extracting explanations from an input word and a semantic resource.

Data: Explanation e_1, Explanation e_2
Result: A similarity reasoning instance.
$p - vectors_1 = \text{getPMIValues}(e_1)$;
$p - vectors_2 = \text{getPMIValues}(e_2)$;
for *each property vector pw_1 (of property p) in $p - vectors_1$* **do**
    if *$p - vectors_2$ has property vector pw_2 of the same property p* **then**
        $matching_{functionalities} = \text{match}(pw_1, pw_2)$;
        if *$matching_{functionalities}$ is not empty* **then**
            $\text{instance.add}(p, matching_{functionalities})$;
        end
    end
end
return the instance;
Algorithm 3. Method *getReasoning* for comparing explanations of two words, returning a similarity reasoning instance.

3.5 Experimentation

In this section, we present a running example to show the richness of a semantic reasoning process compared with a standard model of numeric semantic similarity. In particular, we selected a set of word pairs from a manually-annotated dataset with a various degree of similarity. The goal was twofold: (1) to evaluate the ability to identify the key semantic points (similarity reasoning) of highly-similar annotated word pairs; (2) to evaluate the ability to identify plausible semantic similarity explanations even in case of uncorrelated word pairs. Table 1 shows some examples of similarity reasoning instances[3].

---

[3] The complete set of similarity reasoning instances which have been created from the similarity dataset will be made publicly available in case of acceptance.

Table 1. Examples of Semantic Reasoning instances, using input word pairs from the SimLex-999 similarity dataset.

w_1	w_2	sim	r_w1	r_w2	P	F
spoon	cup	2.02	IsA	IsA	IsA-container	UsedFor-eat(0.69)
apple	juice	2.88	MadeOf	IsA	hasproperty-tasty hasproperty-goodforhealth hasproperty-sweet hasproperty-wet hasproperty-refresh...	createdby-crushapple(1.0) notcapableof-comefromorange(0.71)...
room	bath	3.33	HasA	relatedTo	HasA-door HasA-wall HasA-window	usedfor-comfort(1.0) usedfor-decorate(1.0) usedfor-sleep(1.0) usedfor-livein(1.0) usedfor-sitonit(1.0)
gun	cannon	5.68	RelatedTo	RelatedTo	hasproperty-developin20thcentury partof-weaponry partof-war hasproperty-onekindoffirearm hasproperty-verydangerous...	usedfor-destroything(1.0) usedfor-war(0.53) usedfor-kill(0.39) usedfor-shootthing(0.66) usedfor-commitcrime(0.72) usedfor-violence(0.72)...
prince	king	5.92	relatedTo	relatedTo	hasa-queen definedas-rulerofcountry nothasproperty-elect hasa-army hasa-nopowerindemocracy hasproperty-royal definedas-headofmonarchy	usedfor-rulekingdom(0.72) capableof-rulegroupofperson(1.0) capableof-liveincastle(0.48) usedfor-rulemonarchy(1.0) usedfor-continuedynasty(1.0) usedfor-makedecision(0.87) usedfor-ceremonialfunction(1.0)...
mob	crowd	7.85	RelatedTo	RelatedTo	hasa-identityelement hasproperty-likeclub hasproperty-fun hasproperty-dangerous...	capableof-reachagreement(0.99) capableof-includeindividual(0.99) [usedfor-fun(0.70) usedfor-play(0.85)...

4   Further Considerations on the Impact of the Proposal

In the context of Word Sense Disambiguation, an explanation-based similarity may constitute a novel kind of approach where words in a specific context could match with word senses through the use of correlation between semantic explanations rather than overlapping of word profiles (or vectors). In the Information Retrieval field (IR), complex queries may be seen as sets of shared explanations among the keywords in the query, possibly improving both Precision and Recall. In other words, instead of treating a query as a bag of words, it will be transformed into the explanations obtained by the proposed semantic reasoning. For example, let us consider the 3-keywords query *wolf dog behaviours*. The word *dog* cannot be considered in the role of a pet, so results concerning pets (and so related to cats and parrots) are out of the scope of the query. In a sense, the aim of this proposal would be the removal of unnecessary senses related to the used words by shifting the analysis from a lexical to a semantic basis of correlations. Syntactic parsing is a procedure that often requires semantic information. A semantic reasoning approach could alleviate ambiguity problems at syntactic level by using explanations. For example, major problems for syntactic parsing are prepositional-phrase and verbal attachments. Finally, this model could also help improve the state of the art on Natural Language Generation (NLG), and Summarization, since similarity reasoning can output lexical items which can be also not correlated to the used words in general, but that can play a requested role in a specific linguistic construction.

5 Conclusions and Future Works

In this paper, we proposed a change of paradigm in the context of the research on Computational Linguistics and semantic similarity. In particular, we highlighted the limits of a numeric-based notion of similarity proposing the novel concept of *similarity reasoning*, which is directly inspired from several linguistic and cognitive theories and inherits the numerical nature of a similarity calculation while producing explanations which can be further analyzed to contextualize lexical comparisons. A set of experiments on a manually-annotated dataset of similar (and non-similar) word pairs demonstrated the validity of the approach, and the impact that may have on several tasks. In future work, we aim at extending our idea to lexical entities of higher granularity (such as n-grams, sentences, etc.), through other recently-published annotated data such as the MEN dataset [1] and the Blue Norwegian Parrot dataset [9].

References

1. Bruni, E., Tran, N.K., Baroni, M.: Multimodal distributional semantics. J. Artif. Intell. Res. (JAIR) **49**, 1–47 (2014)
2. Chaffin, R., Herrmann, D.J.: The similarity and diversity of semantic relations. Mem. Cognit. **12**(2), 134–141 (1984)
3. Dumais, S.T.: Latent semantic analysis. Ann. Rev. Inf. Sci. Technol. **38**(1), 188–230 (2004)
4. Finkelstein, L., Gabrilovich, E., Matias, Y., Rivlin, E., Solan, Z., Wolfman, G., Ruppin, E.: Placing search in context: the concept revisited. In: Proceedings of the 10th International Conference on World Wide Web, pp. 406–414. ACM (2001)
5. Gibson, J.J.: The Theory of Affordances. Lawrence Erlbaum, Hillsdale (1977)
6. Heit, E., Rubinstein, J.: Similarity and property effects in inductive reasoning. J. Exp. Psychol. Learn. Mem. Cognit. **20**(2), 411 (1994)
7. Hill, F., Reichart, R., Korhonen, A.: Simlex-999: Evaluating semantic models with (genuine) similarity estimation. arXiv preprint arXiv:1408.3456 (2014)
8. Köhler, W.: Gestalt Psychology. H. Liveright, New York (1929)
9. Kruszewski, G., Baroni, M.: Dead parrots make bad pets: exploring modifier effects in noun phrases. In: Lexical and Computational Semantics (*SEM 2014), p. 171 (2014)
10. Lewis, M., Steedman, M.: Combining distributional and logical semantics. Trans. Assoc. Comput. Linguist. **1**, 179–192 (2013)

11. Manning, C.D., Schütze, H.: Foundations of Statistical Natural Language Processing. MIT Press, Cambridge (1999)
12. Medin, D.L., Goldstone, R.L., Gentner, D.: Respects for similarity. Psychol. Rev. **100**(2), 254 (1993)
13. Miller, G.A.: Wordnet: a lexical database for English. Commun. ACM **38**(11), 39–41 (1995)
14. Murphy, G.L., Medin, D.L.: The role of theories in conceptual coherence. Psychol. Rev. **92**(3), 289 (1985)
15. Murphy, M.L.: Semantic Relations and the Lexicon. Cambridge University Press, Cambridge (2003)
16. Quine, W.V.: Natural kinds. In: Rescher, N. (ed.) Essays in Honor of Carl G. Hempel, pp. 5–23. Springer, Heidelberg (1969)
17. Salton, G., Wong, A., Yang, C.S.: A vector space model for automatic indexing. Commun. ACM **18**(11), 613–620 (1975). http://doi.acm.org/10.1145/361219. 361220
18. Saussure, F.D.: Course in General Linguistics. Duckworth, London (1983). trans. R. Harris
19. Schuler, K.K.: Verbnet: A Broad-coverage, Comprehensive Verb Lexicon. Ph.D. thesis, University of Pennsylvania, Philadelphia, PA, USA (2005). ISBN: 0-542-20049-X, AAI3179808
20. Speer, R., Havasi, C.: Representing general relational knowledge in conceptnet 5. In: LREC, pp. 3679–3686 (2012)

Adopting Semantic Similarity for Utterance Candidates Discovery from Human-to-Human Dialogue Corpus

Roman Y. Shtykh[1,2(✉)] and Mitsuharu Makita[1]

[1] CyberAgent, Inc., Akihabara Dai Bld., Sotokanda 1-18-13, Chiyoda-ku,
Tokyo 101-8608, Japan
{shtykh_roman,makita_mitsuharu}@cyberagent.co.jp
[2] Media Research Institute, Waseda University, 2-579-15 Mikajima,
Tokorozawa 359-1192, Japan

Abstract. Having appropriate utterances in response to user input is an essential element to sustain the flow of conversation in dialogue systems, and a basic and fundamental element for maintaining such conversation coherence is an adjacency pair. To find appropriate candidates for adjacency pairs completion, and thus contribute to avoiding conversational disrupt in casual chatbot systems, we suggest an approach that utilizes human-to-human chat logs, and combines standard Information Retrieval methods and semantic similarity measures based on distributed word representations. The experimental results show the approach improves the quality of utterance pairs compared to standard IR-based methods.

Keywords: Semantic similarity · Utterance candidates · Continuous word embeddings · Word2vec · Human-to-human dialogue corpus

1 Introduction

Typically, automatically producing dialogue utterances is achieved by generating them according to various models or by extracting them from a dialogue logs. While the former approach can be efficient for specific tasks and domains, especially in task-oriented systems, it is costly since it requires creating rules, annotations and other processing methods. The latter approach can be effective and rather inexpensive way for building casual dialogue systems. Although limited to utterances observed in human-to-human dialogues, reusing utterances in a corpus-based approach is very promising [1] greatly reducing a cost needed when generation approach is applied to automatically producing dialogue utterances. This becomes particularly important nowadays for dialogue services with huge volumes of data we can learn from. And this is the approach we choose to build a non-task-oriented chat system.

If we simplistically think about a dialogue structure as a stack of adjacency pairs [2] that are sequences of two related utterances by two different speakers

© Springer International Publishing Switzerland 2016
J.F. Quesada et al. (Eds.): FETLT 2015, LNAI 9577, pp. 139–148, 2016.
DOI: 10.1007/978-3-319-33500-1_12

(as shown in Table 1), to produce utterances that stay in the scope of what is being discussed, two types of coherence have to be considered: coherence inside a newly-created adjacency pairs (`horizontal coherence`) and coherence between neighboring pairs (`vertical coherence`). Horizontal coherence is straightforward in terms the response utterance has to correlate to the first-pair utterance, while vertical coherence depends more on the context and situation. In this paper, we focus on horizontal coherence and discuss an approach to achieve it by finding appropriate candidate utterances for adjacency pair completion from human-to-human dialogues and apply them to human-to-bot dialogues. As we show in this work, utilizing standard Information Retrieval (IR) techniques is not always sufficient for discovered pairs to be truly adjacent, as not every second-pair part candidate (response of the second speaker) can be relevant even it is considered so from IR perspective. We combine such techniques with a semantic similarity model trained on an unannotated human-to-human dialogue corpus to improve the quality of candidate utterances in terms of retrieval precision and increase the discovery rate of relevant high quality candidates. This approach contributes to creating new chatbot systems at low cost in any new domain where big collections of data (logs) are available.

Table 1. Snippet from "Pride and Prejudice" by Jane Austen with (u1, u2) and (u3, u4) adjacency pairs

> u1 "What is his name?"
>
> u2 "Bingley."
>
> u3 "Is he married or single?"
>
> u4 "Oh! Single, my dear, to be sure! A single man of large fortune; four or five thousand a year. What a fine thing for our girls!"

2   Related Works

With the increasing availability of human dialogue data (or data that can be considered as dialogue pairs), such as discussion forums and microblogging services as Twitter, many have chosen the corpus-based (data-driven, or selection) approach for dialogue system construction, e.g., [3–5].

In [6] Twitter data is utilized for response utterance selection and a real-time crowdsourcing is proposed when no appropriate utterance is found in the utterance pair data store. The authors of [7] use human-to-human interaction data to find appropriate utterances in several ways – starting from a random approach, incorporating a two-turn local context and a global context. In our work we use preprocessed[1] adjacency pairs from chat logs as a source for utterance candidates, which can be considered as one-turn local context, and apply

---

[1] As described in Sect. 3.

a semantic model learned from chat logs both for preprocessing and candidate selection. [8]'s approach is similar to ours in terms it thoroughly considers corpus preprocessing steps and apply WordNet-based syntactic-semantic similarity for response generation, and shows good results particularly in TF-IDF based cosine similarity retrieval. In our research, for semantic similarity discovery we use word2vec [9,10] that does not rely on ontologies created by specialists and can be easily trained for various domains. It captures semantic relations among words in dialogue collections and outperforms traditional semantic models [11]. We combine it with standard (TF-IDF) IR based methods to produce better results than when using such methods without semantic similarity applied.

3   Finding Candidate Utterances for Adjacency Pair Completion

As we mentioned in Introduction, the scope of this work is to construct coherent adjacency pairs for a smooth and natural dialogues in a chatbot system. i.e., given an input utterance from a human, we need to provide a response utterance that together with the input utterance forms a natural adjacency pair. Having a history log of human-to-human dialogues, we have to find best candidates for pair completion. The proposed approach to achieve it consists of two steps:

- corpus construction
- utterance selection

In corpus construction step we create utterance pairs from dialogue logs. As our experiments show, this step can be just as simple as pairing neighboring utterances from logs. Although this may be sufficient to obtain quality pairs, other preprocessing steps may be considered to optimize the corpus. For instance, often logs have lengthy 'multi-line utterances' of one user that can be on several topics and contain extraneous utterances until a turn in a conversation is passed to another user. Therefore finer utterance selection when forming pairs has to be considered. In attempt to obtain coherent utterances, we use the following heuristics – we scan the lengthy set of utterances of a user (in a similar manner to TextTiling [12]) starting from the last utterance backwards cutting off the top utterances when they are found to be unrelated to those found at the scanning's start point. Further, to improve their quality, pairs can be checked for their semantic similarity using a model built with word2vec trained on dialogue logs and discard pairs which are below a certain threshold.

After the pair formation phase, the corpus is indexed in such a way that the first part of an utterance pair can be searched and output corresponding responses as candidate utterances from which the selection is done. This is rather a conventional approach for corpus-based dialogue systems and it shows decent results. However, it considers mostly the terms that utterances are comprised of. It ignores semantic correlation between a pair of utterances which is important for constructing natural adjacency pairs, which reflect the mutual perspective of dialogue participants. To increase the quality of utterances at the top of the retrieved candidates' list, we use word2vec (with skip-gram model) for a semantic similarity measure. By Distributional Hypothesis [13], words that occur in similar contexts are likely to have similar meanings. word2vec learns continuous word embeddings from text data and assumes that spatially-close words are similar. By applying the algorithm to the results retrieved through a search engine, we manage to significantly improve the chances of discovered candidates to form a quality adjacency pair, and, as a result, enhance horizontal coherence of the whole dialogue.

4   Evaluation

4.1   Experimental Setup

The experimental data set consists of two parts – an utterance that is submitted to the system, and candidate utterances (potential responses from the chat bot) sampled from chat system logs and manually labeled by their relevancy to the submitted utterance. Since a corpus-based approach requires some volume of utterances of the domain it is applied to (normally, the more, the better), to ensure a rich variety of pair combinations to be selected from and obtained a robust model for similarity measurement with word2vec, we consider utterances corresponding to the most popular topics discovered from the logs by LDA (latent Dirichlet allocation) [14]. We have 12 topics such as weather, dating, marriage, family, job, commuting, online shopping, etc., and four sets for each topic that gives us 48 input utterances in total.

Also, since dialogues vary in the number of participants and to find utterance pairs in multi-user dialogues requires an extra processing step, which is not related to the objectives of this particular research, for corpus construction we consider only two-person dialogues (25 % of all dialogues in the human conversation logs in Japanese from one of CyberAgent[2] chat services).

To verify the validity of the proposed approach, we perform experiments on three indexed data sets which differ by corpus construction methods we discussed in the previous section:

- [basic], where utterance pairs from the logs are used as candidate utterances as-is
- [basic + rules], where simple heuristic rules are applied to utterance pairs as a data cleansing means
- [sim + rules], where the above-mentioned rules are combined with semantic similarity threshold applied to utterance pairs to be indexed in order to make sure we have appropriate adjacency pairs

Input utterance is matched with the first part of adjacency pairs in the search indices using standard IR techniques[3] to come up with response candidates (-sim method in Table 1). Also, we apply semantic similarity analysis[4] (+sim method

---

[2] CyberAgent, Inc. https://www.cyberagent.co.jp/en/.

[3] Apache Lucene is used for the experiments.

[4] https://pypi.python.org/pypi/gensim implementation.

in Table 1) with the model trained on three-month dialogue data (approximately 2 million two-person dialogues) to all candidates utterances retrieved by search, and rerank them by `input - candidate` similarity scores. The settings of word2vec learning are as follows: the size of feature vectors is 200, context window size is 5, and minimum frequency of a word in the corpus is 20. To obtain the phrase representations from word representations, we use a weighted average of all the words in an utterance (except stop words removed as a preprocessing step).

As a result, we have six experiments (two retrieval methods per each index) as shown in Table 2.

Table 2. Experiments combining corpus construction methods (3 types) w/ or w/o semantic similarity analysis applied

Index type	[basic]		[basic + rules]		[sim + rules]	
Sem. reranking	−sim	+sim	−sim	+sim	−sim	+sim

`[basic]`-sim can be considered as the simplest approach for producing candidate utterances for a chat system in a corpus-based approach. It does not require any complex preprocessing, and constructing an adjacency pairs, where its first part is a target for indexing and search, and the second part is a potential response, is sufficient. The similar approach is taken by [6][5] and serves as our baseline.

4.2   Results

Through the experiments described in the previous subsection we have found that applying semantic similarity to the candidate list retrieved by search outperforms standard search alone whatever preprocessing method for corpus construction is chosen. Here we discuss the results in detail.

Since we are interested in the quality of candidate utterances that can be used as responses in the chat, and need to know how well the system brings appropriate candidates to the top of the recommended utterance list, we perform tests with average precision over first n candidate utterances in the list (AP@n), Mean Average Precision (MAP) and Mean Reciprocal Rank (MRR) measures.

First, we look at the percentage of quality utterances in top *10* candidates (Fig. 1). The results are significantly better (as our t-tests with $p < 0.05$ indicate) when the semantic similarity is applied, whichever corpus construction method we use. It also gives us insights on how many candidates worth to be considered as responses provided to the user – for instance, we can randomize top three candidates that look particularly good (see Table 3 for an example ouput).

---

[5] We don't consider an extra crowdsourcing step proposed in the paper though.

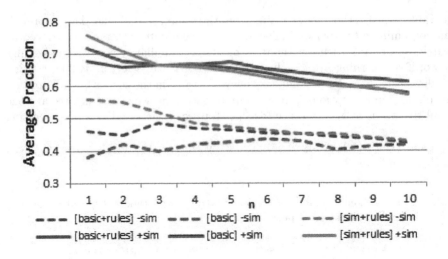

Fig. 1. Average precision

The MAP results for each proposed index and search method combination is shown in Fig. 2. We can see that applying semantic similarity to corpus construction for excluding dialogue pairs which have weak relationships with each other and relying only on text retrieval methods ([sim + rules]-sim) is not much effective compared to the case when semantic similarity analysis is applied to raw adjacency pairs extracted from chat logs ([basic] + sim). Whatever preprocessing method for corpus construction is chosen, it is outperformed by applying semantic similarity to the candidate list retrieved by search.

Finally, since normally we want to limit the number of candidate utterances, we need to know how early a relevant utterance appears in the candidate list. For this, we calculate MRR (Fig. 3). And again, the results are significantly better

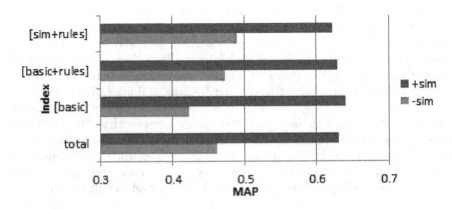

Fig. 2. Mean average precision

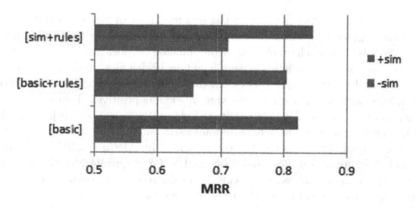

Fig. 3. Mean reciprocal rank

Table 3. Test result examples (top three candidates)

	Input
	明日はクリスマスイブだね^_^
	(Tomorrow is Christmas Eve, isn't it?)
Judgement	**Candidates without sem. similarity applied**
△	うん
	(Yup)
×	今週いっぱい
	(Till the end the this week)
△	そうだね
	(Well...)
Judgement	**Candidates with sem. similarity applied**
○	クリスマスイブなんか知らね
	(I don't care about Christmas Eve.)
○	メリークリスマス!楽しいイブの夜を過ごしてね
	(Merry Christmas! Have a great Christmas Eve!)
○	メリークリスマス!楽しいクリスマスを過ごしてね♪
	(Merry Christmas! Have a joyful Christmas!)

when the semantic similarity is applied, whichever corpus construction method is used.

Also it is noteworthy to mention that the experimental data set contains a chunk of backchannel response utterances having little intrinsic meaning, and they were labeled as negatives in the above-described experiments. But, at the same time, such utterances are also the inherent part of any conversation and cannot be completely ignored. Therefore we also performed the above-mentioned experiments with backchannel expressions labeled as positives. We have found that the results were statistically significant in both cases, with no impact on +sim tests and a slight gain of less than 0.1 on average for -sim tests for those cases where backchannel expressions were labeled as positives. Table 3 shows an extract from one of the tests we conducted w/ and w/o the semantic similarity measure applied and includes backchannel expressions. Here we have an input utterance and top three (as can be recommended by AP@n results) candidates. ○ stands for an appropriate candidate, and × for inappropriate. Backchannel response utterances (marked as △) can be considered as appropriate or inappropriate responses, depending on the user's expectations.

Finally, as we can see from the results, applying heuristic rules and semantic similarity threshold to dialogue pairs when building the corpus produces only a slight improvement to the quality of utterance candidates. However, this combination can be considered as a means to reduce noise in the corpus, and as a result decrease storage costs and potentially increase search speed for large dialogue corpora. For instance, in our experiments, applying the heuristics reduces the index size by 25 %.

5   Discussion and Future Works

Applying semantic similarity to the candidates retrieved by search contributes greatly to response utterance quality, and thus raising chances for adjacency pair completion, in dialogue systems based on human-to-human conversation history. In our experiments, this approach outperforms methods that use standard IR techniques, regardless of which corpus preprocessing steps are chosen. When an appropriate utterance is found, it is very natural.

However, sometimes even if two utterances are found to be semantically similar, a proper adjacency pair may not be formed yet without considering the global context, which has to be taken into consideration for natural dialogue flow generation. For instance, though the parts of the pairs in Table 4 are semantically related, it depends on the context if they can form adjacency pairs.

The first pair of the first example is semantically related and can be considered as a perfect adjacency pair, but it has a time dependency – it makes sense only if the first part of the pair is spoken on Friday. In the second example we show two possible response utterances depending on seasonal context – either of them can be considered as a perfect candidate to compose the adjacency pair depending on whether the first part is spoken in cold or warm season. This is what cannot be captured without considering the contextual semantics of the dialogue, and although it is out of scope of the presented work it shows the direction for our further research for ensuring the vertical coherence of a dialogue.

The proposed method is likely to be effective when large domain data logs of human-to-human dialogues are available. For the experiments we have used a

Table 4. Context-dependent pairs

Example 1	
Pair part 1	明日お仕事ですよね。
	(Tomorrow is a working day, isn't it?)
Pair part 2	土日は仕事だよ
	(Yes, I work on weekends.)
Example 2	
Pair part 1	最高気温6度だったから
	(The maximum temperature was 6 degrees Celsius.)
Pair part 2 (1)	おおおさむうですね
	(Oh, so cold.)
Pair part 2 (2)	あったけーーーおう
	(So warm!)

large volume of a popular chat service (three-month data), and further increasing the data volume has the potential of improving the quality of utterance candidates.

Finally, the proposed approach compares terms found in utterance pairs without considering their order. We are in process of experimenting with substituting word vectors with more elaborate approaches, such as Paragraph Vector [15], which may introduce better contextual cohesion of terms, thus improving chances to discover high quality adjacency pairs.

References

1. Gandhe, S., Traum, D.: I've Said It Before, and I'll Say It Again: an empirical investigation of the upper bound of the selection approach to dialogue. In: Proceedings of the SIGdial 2010 Conference, Tokyo, Japan, pp. 245–248 (2010)
2. Schegloff, E.A.: Sequence Organization in Interaction: A Primer in Conversation Analysis I. Cambridge University Press, Cambridge (2007)
3. Huang, J., Zhou, M., Yang, D.: Extracting chatbot knowledge from online discussion forums. In: Proceedings of the 20th International Joint Conference on Artifical Intelligence, Hyderabad, India, pp. 423–428 (2007)
4. Wu, Y., Wang, G., Li, W., Li, Z.: Automatic chatbot knowledge acquisition from online forum via rough set and ensemble learning. In: Proceedings of the IFIP International Conference on Network and Parallel Computing, Shanghai, China, pp. 242–246 (2008)
5. Higashinaka, R., Kobayashi, N., Hirano, T., Miyazaki, C., Meguro, T., Makino, T., Matsuo, Y.: Syntactic filtering and content-based retrieval of Twitter sentences for the generation of system utterances in dialogue systems. In: Proceedings of International Workshop Series on Spoken Dialogue Systems Technology, Napa, USA, pp. 113–123 (2014)

6. Bessho, F., Harada, T., Kuniyoshi, Y.: Dialog system using real-time crowdsourcing and Twitter large-scale corpus. In: Proceedings of the SIGdial 2012 Conference, Stroudsburg, PA, USA, pp. 227–231 (2012)
7. Gandhe, S., Traum, D.: Creating spoken dialogue characters from corpora without annotations. In: Proceedings of Interspeech-2007, Antwerp, Belgium, pp. 2201–2204 (2007)
8. Nio, L., Sakti, S., Neubig, G., Toda, T., Nakamura, S.: Utilizing human-to-human conversation examples for a multi domain chat-oriented dialog system. IEICE Trans. **97–D**(6), 1497–1505 (2014)
9. Mikolov, T., Chen, K., Corrado, G., Dean, J.: Efficient estimation of word representations in vector space. In: Proceedings of ICLR Workshop (2013)
10. Mikolov, T., Sutskever, I., Chen, K., Corrado, G., Dean, J.: Distributed representations of words and phrases and their compositionality. In: Advances in Neural Information Processing Systems, pp. 3111–3119 (2013)
11. Baroni, M., Dinu, G., Kruszewski, G.: A systematic comparison of context-counting vs. context-predicting semantic vectors. In: Proceedings of the 52nd Annual Meeting of the Association for Computational Linguistics, Baltimore, Maryland, USA, pp. 238–247 (2014)
12. Hearst, M.A.: TextTiling: segmenting text into multi-paragraph subtopic passages. Comput. Linguist. **23**(1), 33–64 (1997)
13. Firth, J.R.: A synopsis of linguistic theory 1930–1955. Studies in Linguistic Analysis, pp. 1–32 (1957)
14. Blei, D., Ng, A., Jordan, M.: Latent Dirichlet allocation. J. Mach. Learn. Res. **3**, 993–1022 (2003)
15. Le, Q., Mikolov, T.: Distributed representations of sentences and documents. In: Proceedings of the 31st International Conference on Machine Learning (ICML-2014), pp. 1188–1196 (2014)

Modelling Goal Modifications in User Simulation

Stefan Hillmann$^{(\boxtimes)}$ and Klaus-Peter Engelbrecht

Quality and Usability Lab, Telekom Innovation Laboratories,
Technische Universität Berlin, Berlin, Germany
stefan.hillmann@tu-berlin.de, klaus-peter.engelbrecht@telekom.de

Abstract. User simulation is frequently used to evaluate spoken dialogue systems. Previous work in this field primarily focused on the users' interaction behaviour. Less attention has been paid to the users' goals, how they relate to the system capabilities, and how they may change over the course of a dialogue. User goals may be underspecified or overspecified in comparison to the attributes the system uses to describe objects in its domain. Goal modifications can occur, e.g., if no database entry matches the user query. Analysing empirical data, we show that the definition of possible goals and goal modifications impacts the results of a user test significantly in terms of system performance and discovered usability problems. We propose a task modelling approach able to represent such variable goals in user simulations. Dialogues simulated using this approach are shown to be more similar to empirical data than dialogues simulated with conventional task models.

1 Introduction

Due to advances in language technology, spoken dialogue systems (SDS) are becoming more complex, requiring more careful testing of individual components and the integrated system.

Dialogue systems are often essentially information systems, where the user inputs values (or *constraints*) for several query attributes, and the system looks up matching entries in a database. In a complex domain, the query attributes used by the system may not be known to the user. Furthermore, some attributes used by the system may not be relevant, or relevant attributes may not be supported by the system (i.e. the task is underspecified or overspecified). Moreover, users will not just pursue a fixed goal, but will trade off goals against options. For example, the user may be willing to modify one of the query attributes, if the system can not provide a matching result.

In user tests, users usually get a task (i.e. a goal provided by the experimenter) which they have to solve with the system. The tasks have to be carefully selected for a test, as the selection can significantly impact the outcome of an evaluation. Furthermore, if the tasks do not cover the right system functionalities, some usability-problems may not be discovered.

User tests are expensive and not necessary if the goal of the evaluation is to test for errors in the implemented dialogue strategy (rather than the users'

© Springer International Publishing Switzerland 2016
J.F. Quesada et al. (Eds.): FETLT 2015, LNAI 9577, pp. 149–159, 2016.
DOI: 10.1007/978-3-319-33500-1_13

response to the strategy). In such cases, a more economical test procedure is to use user simulation to test the system against known properties of the users quickly and cheaply [1–6]. Alternatively, simulations may be used to optimize the parameters of a statistical dialogue system (e.g., [7–9]).

In previous work on user simulation, task modelling has usually merely been discussed as a side-topic. In early simulation approaches [7] a task model is missing completely, while the modelling effort went into selecting appropriate types of dialogue acts given a system turn. Task models were soon introduced, but still very simple, consisting in a fixed value for each attribute the system may ask for. Examples are slots of a slot filling system or fields in VoiceXML scripts (e.g., [3,9,10]). Again, much effort went into the selection of appropriate speech acts, the number of constraints contained in a user dialogue act, or the order in which constraints are conveyed to the system [8,9].

A few works did consider more accurate task models. Pietquin [8] modelled preferences regarding the constraints in order to determine how likely each constraint is relaxed. However, the new constraint would always be determined by the system rather than the user. Chung [1] assumed that users may change their goals, but only simulated random goal modifications. The agenda-based user model [11] also foresaw goal modifications, however this part of the model was later sacrificed for a tractable training procedure.

In this paper, we argue that user simulations should use more sophisticated task models in order to mimic real user dialogues more accurately and more completely. A more accurate model will yield better performance predictions, and a more complete model will yield a higher number of detected errors due to more substantially different and valid test cases. We propose to model underspecified and overspecified tasks, as well as goal modifications users make during the interaction, as explained in Sect. 2. The impact of such properties of the user goal on test results is illustrated in Sect. 3, using data of a real user test with a restaurant information system. Afterwards, a task modelling approach implementing the proposed properties of goals is presented. We show that the new model leads to simulation results which are significantly more similar to those of a subject-based test than previous approaches.

2   Task Modification Classification

If we assume that possible values of a constraint can be structured in a taxonomy (as illustrated in Fig. 1) three types of task modifications can be distinguished: *Expansion*, *Refinement* and *Exchange*. Expansion and Refinement are similar to the approach of constraint relaxation described in [12].

Expansion means that a concept is changed to a parent concept or even a parent of a parent (i.e. a more general one). In the example shown in Fig. 1, *cuisine=italian* can be changed to *cuisine=mediterranean* (short: *italian →
mediterranean*). Refinement describes the reversed case. Here, the new value is a child (or a child's child) of the old value (e.g., *mediterranean → spanish* in Fig. 1). The third case, Exchange, covers all modifications of a value that are

not an Expansion or a Refinement. In Fig. 1, the following modifications are examples for an Exchange: *italian* → *spanish, italian* → *chinese* or *spanish* → *asian.*

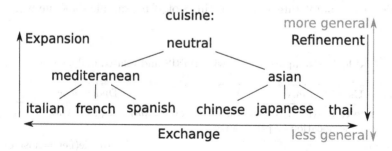

Fig. 1. A taxonomy of exemplary values for the concept *cuisine* and the relations between them. In this representation, a child (e.g. *italian*) is less general than its parent (i.e. *mediteranean*).

A modification can be initiated by either the system or the user [13, pp. 396–397]. A system-initiated modification can be triggered by a system request to expand, refine or exchange a constraint. In each of these cases the system can utter an open-ended prompt, suggest to change a specific attribute-value-pair (AVP) or propose a concrete new value for an attribute (e.g., *average* instead of *cheap* for the price). User-initiated modifications can occur because the system did not recognize the previously uttered value, leading to the user's believe that this concept is not covered by the system domain. Similarly, a user may accept a confirmation request for a misunderstood constraint. Furthermore, the user can spontaneously and by intuition modify an attribute, e.g.:

- if the system asks for an attribute that was not specified in the task description for the experiment,
- if there is a mismatch between the user's and the system's domain model, (e.g., the user says "10 Euro" for "cheap" but the system maps this to "average", or the user says "Chinese" while the task indicates "duck" as food type)
- or if the user deviates from the task, e.g., by assuming that if the task description specifies no time, this means "now".

The cases of mismatches and deviations were not mentioned in the previous literature but are included here to describe all modifications observed in the database.

3   Empirical Analysis

We analysed data collected with the BoRIS SDS [13, pp. 241–244]. BoRIS allows its users to find a restaurant in Bochum (Germany). In a mixed initiative dialogue, it collects constraints until a set of three or less matching restaurants has been found in a database, or until all constraints the system can handle are set. If no restaurant is found, the system offers the user to modify constraints. On the other hand, if more than three restaurants are found, the system offers the user to refine constraints. As a further feature, the user can name a slot, and by this trigger a prompt targeted to fill this slot. An example dialogue is given in Table 1.

Table 1. Example dialogue with BoRIS and related dialogue acts.

System or User utterance	Dialogue act
S1: Welcome to BoRIS... You can query for a date, day time, cuisine, location or price range	request(ALL)
U1: Cuisine	provide(field=cuisine)
S2: You can choose between German, French, ...	indicateValues(cuisine)
U2: Id like french food	provide(cuisine="french")
...	...
S3: Which price range?	request(price)
U3: Medium	provide(price="mid")
S4: I could not find a restaurant matching your query. Would you like to change your query?	offerModification()
U4: Yes	accept()
S5: You can change date, day time, cuisine, location or price range.	request(field)
U5: Italian	provide(cuisine="italian")
S6: The possible restaurants are: name, address	inform()

The experiment's details are described in [13, pp. 237–311], and the following gives just a short summary. Forty users performed five tasks each, resulting in 200 dialogues (i.e. 2003 turns). Four tasks were predefined by the experimenter, and the fifth task was defined by the user before that trial. Since consistent task definitions were needed for our current data analysis, only the four predefined tasks were used.

Special care was taken for the users to behave naturally and in a variable way. Therefore, the predefined tasks were partly described as graphical scenarios [13, p. 134] to avoid priming effects, and some included the specification of a constraint to modify the query, if no or too many restaurants were found. In this case, either a new constraint, or just the attribute to change was specified. In addition, some tasks were underspecified, i.e. not for all system slots a desired value was given. For example, one task was to find a restaurant that serves duck. For other attributes, like the price range or the location of the restaurant, the user could either invent constraints if she felt this was necessary, or she could set this attribute to "neutral".

3.1 Task Modifications in the BoRIS Database

Table 2 shows how often each type of task modification occurred in the empirical data (overall 288 cases). In 32 % of the cases, the modification was initiated by a system request. Most of the time, the system asked for an exchange rather than a refinement. In 5.5 % of the cases, the user changed a constraint apparently because she thought the system could not understand it. Only two times, a misunderstood constraint was affirmed. In another third (35.5 %) of the cases, users invented a value for an attribute not specified in the task when the system asked for it. Additionally, 21.5 % of the task modifications were caused by mismatches in the domain models or the user's deviations from the task. Finally, 4.5 % (13 cases) of the task modifications could not be explained by the available data and were classified as unexplained.

Table 2. Frequencies of task modifications.

Cause for modification	Frequency by type			Sum
	Expansion	Refinement	Exchange	
System requested exchange	18	3	71	92
System requested refinement	0	1	0	1
System's language understanding failed	1	3	12	16
User confirms misunderstood concept	0	0	2	2
System asked for unspecified attribute	0	102	0	102
Mismatches in domain models	2	0	31	33
User deviated from task	3	17	9	29
Unexplained	2	2	9	13
Sum	26	128	134	288

3.2 Impact of Task Modifications on the Outcome of a Usability Test

Task modifications had a significant impact on the outcome of the usability test in terms of system performance measurements as well as usability problems found. In sum, 30% of the dialogue turns in the database were concerned with negotiating the user's constraints. In a dialogue, all system and user turns after the system's first request for modification were defined as related to the negotiating process. Thus, we split the complete corpus into two parts at the level of single turns. One part containing all turns which are related to a modification (and its negotiation), and another containing all turns which are not related to a modification. Between these two parts differences (2-sample t-test, $\alpha = 0.01$) were found for critical system performance measures, e.g.:

- *average words per system turn* before ($M = 20.6$, $SD = 7.7$) and after ($M = 18.1$, $SD = 2.9$) the first modification, $t(273) = 2.71$, $p = 0.007$
- *constraints per user turn* before ($M = 1.28$, $SD = 0.58$) and after ($M = 1.0$, $SD = 0.35$) the first modification, $t(235) = 3.74$, $p < 0.001$
- *concept error rate* [13] before ($M = 0.46$, $SD = 0.36$) and after ($M = 0.25$, $SD = 0.24$) the first modification, $t(235) = 4.23$, $p < 0.001$.

Overall, on the level of complete dialogues it was found that dialogues with task modifications lasted significantly longer than other dialogues ($t(156) = 6.55$, $p < 0.01$).

Two of the authors analysed the system log files of the experiment and found 41 dialogue management and prompt wording problems, as described in [14]. 17 problems were related to the task, and 9 were related to task modifications. It should be noted, for the latter 9 problems a user simulator has to support goal modifications in order to make those problems observable in simulated data.

3.3   Underspecified Tasks and Initiative

As stated in the introduction of Sect. 3, some tasks given to the users were underspecified, i.e. not for all system slots a goal value was provided. It is no surprise that the number of constraints uttered by the users in response to open-ended system questions was correlated with the number of constraints provided by the task description (cf. Table 3). Note that technically the user could have set undefined slots to the value "neutral" in such situations. Also, some users refined an undefined attribute to a specific value when it was queried by the system. However, undefined attribute were rarely specified in response to open-ended prompts.

Table 3. Number of constraints specified in the task compared to user initiative in terms of number of used constraints.

	Task			
	1st	2nd	3rd	4th
Number of task constraints in task description	1	5	4	3
Average number of user constraints after open prompts	1.26	3.26	2.59	2.26

4   Implementation

Given that goal modifications or underspecified tasks impact the behaviour of users and the system, a user simulator in which both aspects are modelled should generate more realistic dialogues, than a simulator ignoring them. To test this, we implemented a simulation framework, consisting of a user model, a speech understanding error model and the system. During simulation, the user model and the system communicate using dialogue acts [15, p. 840 ff.]. A dialogue act consists of a type and 0 to n attributes or attribute-value pairs (AVPs; see Table 1 for an illustration). To model speech understanding errors we use the algorithm described in [6] where deletions and insertions are generated based on an AVP confusion matrix. The required probabilities are trained on the real user data. The user model implementation is separated into a task model and an interaction model. The task model defines goals and possible goal modifications, whereas the interaction model defines how users respond to system dialogue acts when pursuing the goal defined in the task model.

4.1 Task Model

The proposed task model is based on the conventional approach where a goal value is provided for each system slot (e.g., [11,14]). Added is support for modelling goal modifications and underspecified tasks.

This is realised in the task model's implementation by an ordered list of constraints. The user may switch to alternative constraints during the interaction in order to perform a task modification. The order of the list defines the user's preferences among the constraints. Section 5 gives further information on the determination of the constraints' order. Given such a list, different types of system-initiated modifications are possible:

- Open-ended question to change any constraint: The goal is updated with the first constraint in the list.
- Offer to modify a specific attribute: If constraints with the required attribute are in the list, the goal is updated with the first of these constraints.
- Proposal to use a specific constraint: If this constraint is in the list, the goal is updated accordingly.

In addition to task modifications it should be possible to model that the user introduces certain constraints only if the system asks for the respective attribute. Therefore, goal constraints are categorized into active and passive constraints. Passive constraints are only uttered in case the system queries the attribute explicitly. Thus, undefined (or "neutral") attributes would usually be specified as passive, whereas constraints defined in the task would be active.

4.2 Interaction Model

The interaction model is used to describe how users interact with the system. Decisions at each turn are based solely on the previous system dialogue act. A mixture of deterministic rules and random selection is used to model the decision process. Given a system dialogue act, the interaction model determines an appropriate user dialogue act to respond with. Depending on the selected dialogue act, AVPs may be included in the user turn to further specify the information sent to the system.

If the system requests a value for one or several specific slots, the interaction model answers with all matching constraints that are specified in the task model. In case of an open-ended question, the number of constraints uttered by the user (n), is sampled from a binomial distribution (as in [16]). Then, n constraints are randomly selected from the active constraints. If the system requests a slot name, as in row S5 in Table 1, two cases are possible. If the system offered a modification previously (e.g., S4 in Table 1) and the interaction model accepted the offer (i.e. could modify a constraint in the task model), the modified constraint is stored and used to answer the slot name request. In the other case, no constraint was previously modified and preselect. Here, a random slot is selected.

If the system offers the user to modify the search query, this is accepted if a constraint is left in the list of acceptable modifications. Otherwise, it is declined. Note that BoRIS always uses the open-ended strategy described in Sect. 4.1. Thus, the user can choose an arbitrary constraint for modification. If a goal modification is performed, this is remembered by the user model in order to continue the dialogue consistently.

Finally, in case of an explicit confirmation request, the user model only affirms if all constraints to be confirmed are correct.

5   Evaluation

We evaluated how modelling goal modifications and passive constraints improves the similarity of the simulated dialogues to the real users' dialogues with BoRIS. The impact of either goal modifications or the active/passive distinction can be assessed independently by running simulations with the following task models:

A: Both, goal modifications and passive constraints are fitted to the empirical data.

B: No goal modifications, passive constraints are fitted to empirical data.

C: Goal modifications are fitted to empirical data, but no passive constraints are used.

D: Goal modifications are random, passive constraints are fitted to empirical data. This model reveals how sensitive it is to model modifications explicitly.

The task model was fitted to the data as follows. The initial goal was obtained by extracting the constraints pursued by the user until the first offer for modification (or until the end of the dialogue if there was no such offer). Constraints were set to passive if they were not mentioned in the task description given to the test users. The ordered list of alternative constraints for task modifications was obtained by extracting all modifications co-occurring with offers for modification by the system. Since the extracted initial goal and the list of alternative constraints could differ between dialogues even if the task description was the same, a task model instance was specified for each dialogue in the corpus.

In the D condition, offers for modifications by the system are accepted at random. In case of acceptance, a random constraint is selected from the system's

Table 4. Evaluation results for simulations using the models A to D.

	Model			
	A	B	C	D
Simulated dialogues	1,580	1,580	1,580	1,580
Common utterances	55	55	59	55
Recall	0.49	0.49	0.52	0.49
Precision	0.60	0.61	0.22	0.37
CvMD	0.062	0.162	0.064	0.074
Worse than A	-	Yes*	No	No

$^{*}p < 0.05$
CvMD = Cramér-von Mises divergence

domain model and added to the modification list. In the condition with no modifications (B), offers for modification are always declined.

In all simulations, the same speech understanding error model and interaction model were used. Each model was used to generate a simulated corpus containing 1,580 dialogues; altogether 6,320 simulated dialogues.

To compare the user turns in a simulated corpus to the BoRIS database, we calculated recall and precision as proposed by Schatzmann et al. [17]. However, while Schatzmann et al. compute recall and precision based on user utterances in a specific dialogue context, we base them on the set of unique user turns in each corpus. Statistical significance of the improvement in performance prediction observed with the fully fitted model compared to models where modifications or passive constraints were random or not modelled at all was determined using the normalised Cramér-von Mises divergence (CvMD) [18]. The CvMD can be used to compute the similarity of two frequency distributions. Using the performance function shown in Eq. 1, we scored single dialogues and collected the frequencies of scores. In Eq. 1, *task success* is either 1 (in case of successful task completion) or 0 (no success), and *no.Turns* the number of turns in a dialogue.

$$Y = 100 \times task\ success - no.Turns \tag{1}$$

Results are shown in Table 4 and described in the following. The row *Common Utterances* in Table 4 gives the number of unique utterances which appear in the BoRIS corpus and the respective simulated corpus. If no goal modifications are simulated (B), the performance predictions are significantly worse than in (A), since modification sub-dialogs are never simulated. Recall and precision are very similar, though.

On the other hand, if no passive constraints are specified (C), the CvMD is roughly equal to A, but many "wrong" utterances are simulated, leading to a much lower precision. Finally, when simulating random goal modifications (D), the CvMD is better (i.e. smaller) than if no modifications are simulated at all (B) and slightly larger than for model A. Unsurprisingly, the simulation used some AVPs in D which never occurred in the tasks for the real users, leading to lower precision. Since with different constraints more or fewer restaurants may be found, this also impacts the performance, leading to a higher CvMD.

The results show that for performance predictions a good model of modifications is important, while for simulating correct utterances, the active-passive distinction is crucial.

6 Discussion and Conclusion

This paper argued that user simulation for testing or training SDSs profits from more precise and sophisticated task models. We presented a modelling approach which can be used with different interaction models. The suggested task model leads to a higher similarity between the empirical and our simulated data.

The presented empirical analysis may be criticized for being based on laboratory data, in which users pursue goals given to them by an experimenter rather than their own goals. However, the laboratory setting was necessary as it determined the user's initial goal. Furthermore, the task descriptions were thoughtfully designed to represent different types of misconceptions users may have about the system or the domain. Indeed, different types of task modifications were observed in the empirical data, of which many were not defined in the task descriptions given to the user.

The proposed task model is a compromise between accuracy and complexity. Ideally, all characteristics of user goals are modelled, including preferences among different values for one attribute and dependencies between the values of different attributes. Unfortunately, this is algorithmically complex, and it would be difficult —if not impossible— to empirically examine goals of real users on this level of detail.

In this paper, we could show that a model which considers goal modifications allows more accurate reproduction of data from empirical studies than models ignoring that aspect. However, the full potential of modelling goals more deeply, including preferences and allowed modifications, lies in the simulation of more sophisticated dialogue strategies for trading off user goals against available database matches. Future work will analyse if the proposed task model is sufficiently detailed to perform such simulations meaningfully. To do so, we will use our simulator to evaluate spoken dialogue systems providing new dialogue strategies for goal modification.

References

1. Chung, G.: Developing a flexible spoken dialog system using simulation. In: Proceedings of the 42nd Annual Meeting on Association for Computational Linguistics, pp. 63–71. Association for Computational Linguistics (2004)
2. López-Cózar, R., Callejas, Z., McTear, M.: Testing the performance of spoken dialogue systems by means of an artificially simulated user. Artif. Intell. Rev. **26**, 291–323 (2006)
3. Ito, A., Shimada, K., Suzuki, M., Makino, S.: A user simulator based on voiceXML for evaluation of spoken dialog systems. In: Proceedings of Interspeech 2006 (2006)

4. Ai, H., Weng, F.: User simulation as testing for spoken dialog systems. In: Proceedings of the 9th SIGdial Workshop on Discourse and Dialogue, pp. 164–171 (2008)
5. Scheffler, T., Roller, R., Reithinger, N.: SpeechEval: a domain-independent user simulation platform for spoken dialog system evaluation. In: Proceedings of IWSDS 2011, pp. 295–300 (2011)
6. Engelbrecht, K.-P., Möller, S.: Correlation between model-based approximations of grounding-related cognition and user judgments. In: Proceedings of Interspeech 2012, pp. 246–249 (2012)
7. Eckert, W., Levin, E., Pieraccini, R.: User modeling for spoken dialogue system evaluation. In: Proceedings of IEEE Automatic Speech Recognition and Understanding Workshop ASRU 1997, pp. 80–87. IEEE (1997)
8. Pietquin, O.: A framework for unsupervised learning of dialogue strategies. Ph.D. thesis (2004)
9. Schatzmann, J., Thomson, B., Weilhammer, K., Ye, H., Young, S.: Agenda-based user simulation for bootstrapping a POMDP dialogue system. In: Human Language Technologies: The Conference of the North American Chapter of the Association for Computational Linguistics (NAACL 2007), (Morristown, NJ, USA), pp. 149–152. Association for Computational Linguistics (2007)
10. Scheffler, K., Young, S.: Corpus-based dialogue simulation for automatic strategy learning and evaluation. In: Proceedings of NAACL Workshop on Adaptation in Dialogue Systems, pp. 64–70 (2001)
11. Schatzmann, J., Young, S.: The hidden agenda user simulation model. IEEE Trans. Audio Speech Lang. Process. **17**, 733–747 (2009)
12. Varges, S., Weng, F., Pon-Barry, H.: Interactive question answering and constraint relaxation in spoken dialogue systems. Nat. Lang. Eng. **15**(01), 9–30 (2009)
13. Möller, S.: Quality of Telephone-Based Spoken Dialogue Systems. Kluwer Academic Publishers, Boston (2005)
14. Engelbrecht, K.-P.: Estimating spoken dialog system quality with user models. Ph.D. thesis, Technische Universität Berlin (2012)
15. Jurafsky, D., Martin, J.H.: Speech and Language Processing, 2nd edn. Pearson, New Jersey (2009)
16. Keizer, S., Gašić, M., Jurčíček, F., Mairesse, F., Thomson, B., Yu, K., Young, S.: Parameter estimation for agenda-based user simulation. In: Proceedings of the 11th Annual Meeting of the Special Interest Group on Discourse and Dialogue, pp. 116–123 (2010)
17. Schatzmann, J., Georgila, K., Young, S.: Quantitative evaluation of user simulation techniques for spoken dialogue systems. In: Proceedings of the 6th SIGdial Workshop on Discourse and Dialogue, pp. 45–54 (2005)
18. Williams, J.D.: Evaluating user simulations with the Cramér-von Mises divergence. Speech Commun. **50**, 829–846 (2008)

Author Index

Printed in the United States
By Bookmasters